STUDENT SOLUTIONS MANUAL
FOR
CHANCE AND CHANGE

3rd EDITION

BY
G. VIGLINO AND E. VIGLINO

Ramapo College of New Jersey
January 2017

CONTENTS

PART 1
PROBABILITY

PART 2
CALCULUS

PREFACE

This manual contains the solutions to all of the odd-numbered Exercises, and to all (including even-numbered) of the Review Exercises for both the Probability and the Calculus Parts of the textbook:

Chance and Change

by Giovanni Viglino and Enrico Viglino

§1.1. Definitions and Examples
Page 9

1. {Jan, Feb, Mar, Apr, May, Jun, Jul, Aug, Sept, Oct, Nov, Dec}

3. {CF, CW, VF, VW, SF, SW}   **5.** {H1, H2, H3, H4, H5, H6, T1, T2, T3, T4, T5, T6}

7. {AA, AB, AC, BA, BB, BC, CA, CB, CC }

9. Calling the ducks: A, B, and C: {ABC, ACB, BAC, BCA, CAB, CBA}

11. Since there are 26 red cards in a deck of 52 cards: $Pr(R) = \dfrac{26}{52}$.

13. Since there are 6 red face cards (J,Q,K of hearts and of diamond) in a deck of 52 cards:

$Pr(\text{Red Face}) = \dfrac{6}{52}$.

15. In the 36-element sample space of Figure 1.2 there are six doubles:

$(1,1), (2,2), (3,3), (4,4), (5,5), (6,6)$. Hence: $Pr(\text{doubles}) = \dfrac{6}{36}$.

17. In the 36-element sample space of Figure 1.2 there are six successes: $(1,5), (2,5), (3,5),$

$(4,5), (5,5), (6,5)$. Hence: $Pr(\text{second die is 5}) = \dfrac{6}{36}$.

19 and 21:
$$\begin{matrix}(1,2)(1,3)(1,4)(1,5)\\(2,1)(2,3)(2,4)(2,5)\\(3,1)(3,2)(3,4)(3,5)\\(4,1)(4,2)(4,3)(4,5)\\(5,1)(5,2)(5,3)(5,4)\end{matrix}$$
note: no doubles

19. The number 5 appears in last row and last columnn. Hence $Pr(\text{5 appears}) = \dfrac{8}{20}$.

21. Since the number 5 cannot occur twice, there are still 8 successes, and $Pr(\text{5 appears}) = \dfrac{8}{20}$.

23 and 25:
$$\begin{matrix}(1,1)(1,2)(1,3)(1,4)(1,5)\\(2,1)(2,2)(2,3)(2,4)(2,5)\\(3,1)(3,2)(3,3)(3,4)(3,5)\\(4,1)(4,2)(4,3)(4,5)(5,5)\\(5,1)(5,2)(5,3)(5,4)(5,5)\end{matrix}$$
note: some doubles

23. A 5 appears in the last row and last column. [Do not count $(5,5)$ twice.] Hence $Pr(\text{5 appears}) = \dfrac{9}{25}$.

25. Removing the element (5,5) from the 8 successes of 23 we have: $Pr(\text{exactly one 5}) = \dfrac{8}{25}$.

27 through 31: $\begin{bmatrix} PN & PD & PQ \\ & ND & NQ \\ & & DQ \end{bmatrix}$

27. Success: DQ. So: $Pr(35) = \dfrac{1}{6}$.

29. Success: PQ, ND, NQ, DQ. So: $Pr(\text{at least } 15) = \dfrac{4}{6}$.

31. Success: PD. So: $Pr(\text{less than } 10) = \dfrac{1}{6}$.

33. Success: PD, PQ, ND, NQ, DQ. So: $PR(>10) = \dfrac{5}{6}$.

35 through 45: $\begin{bmatrix} P_1P_2 & P_1N & P_1D & P_1Q \\ & P_2N & P_2D & P_2Q \\ & & ND & NQ \\ & & & DQ \end{bmatrix}$

35. Success: P_1Q, P_2Q, NQ, DQ. So: $Pr(>25) = \dfrac{4}{10}$.

37. Success: NQ, DQ. So: $Pr(>27) = \dfrac{2}{10}$.

39. Success: P_1P_2. So: $Pr(2) = \dfrac{1}{10}$.

41. Success: $P_1N, P_1D, P_1Q, P_2N, P_2D, P_2Q$. So: $Pr(\text{Exactly one penny}) = \dfrac{6}{10}$.

43. Success: ND, NQ, DQ. SO $Pr(\text{No penny}) = \dfrac{3}{10}$.

45. Success: $P_1D, P_1D, P_2D, P_2Q, ND, NQ, DQ$. So $Pr(>10) = \dfrac{7}{10}$.

47. How the bottom block is situated is irrelevant, since it is equally likely that any of the six colors can appear on the top of the top block. So, the probability hat it is red is $\dfrac{1}{6}$.

49. The probability that the bottom of the two-block stack is **any** one of the six colors is $\dfrac{1}{6}$. In particular, the probability that the bottom of the two-block stack is the particular color at the top of the stack it $\dfrac{1}{6}$.

51. Whatever number you pick, the probability that your friend picks that number is $\dfrac{1}{5}$.

53. $\left\{ \begin{matrix} J & J & M & M & B & B \\ M & B & J & B & J & M \\ B & M & B & J & M & J \end{matrix} \right\}$ $Pr(J \text{ next to } M) = \dfrac{2}{6}$.

55. $\left\{ \begin{matrix} J & J & M & M & B & B \\ M & B & J & B & J & M \\ B & M & B & J & M & J \end{matrix} \right\}$ $Pr(B \text{ first}) = \dfrac{2}{6}$.

57. Since Bobbie hit safely 6 out of 20 times at bat, the probability that he will hit safely the next time at bat is $\dfrac{6}{20} = 0.3$.

59. At this point, Bobbie hit safely 6 out of 21 times at bat. Hence, the probability that he will hit safely the next time at bat is $\dfrac{6}{21} \approx 0.29$.

61. If Bobbie gets hits safely on his 301th time at bat, then he will have 101 hits out of 301. Hence, the probability that he will hit safely the next time at bat is $\dfrac{101}{301} \approx 0.34$.

63.

	Work-Hours per Week (WH)				
	WH < 5	5 ≤ WH < 10	10 ≤ WH < 20	20 ≤ WH	Sum
Freshman	175	210	115	55	555
Sophomore	200	178	95	83	556
Junior	115	205	101	75	496
Senior	250	225	55	50	580
Sum	740	818	366	263	2187

(a) $Pr(\text{Junior}) = \dfrac{496}{2187} \approx 0.23$

(b) $Pr(\text{at least } 20) = \dfrac{263}{2187} \approx 0.12$

(c) $Pr(\geq 10) = \dfrac{366 + 263}{2187} = \dfrac{629}{2187} \approx 0.29$

(d) $Pr(\text{Senior} \geq 20) = \dfrac{50}{2187} \approx 0.023$.

65. (a) Since there are 4 successes (drawing an ace) and $52 - 4 = 48$ failures (not drawing an ace), the odds in favor of drawing an ace are 4 to 52, or 1:12.

(b) Since there are $52 - 12 = 40$ non-face cards, the odds against drawing a face card are 40 to 12, or 10:3.

67. Since the odds in favor of your ship coming in are 3 to 5, your ship will come in 3 out of 8 times. Hence: $Pr(\text{Ship comes in}) = \dfrac{3}{8}$.

69. Since $Pr(\text{position}) = \dfrac{75 \text{ successes}}{100 \text{ possibilities}}$ the odds in against securing an interesting position are $\dfrac{25 \text{ failures}}{75 \text{ successes}}$; which is to say 25 to 75, or 1:3.

§1.2. Unions and Complement of Events
Page 23

1. Twelve of the 52 cards are face cards. Hence $Pr(\text{Face}) = \dfrac{12}{52}$.

3. Since there are $12 + 4 = 16$ face or ace cards, there are $52 - 16 = 36$ cards that are nether a face or ace card. Hence $Pr(\text{Not Face or Ace}) = \dfrac{36}{52}$; or, if you prefer:

$$Pr(\text{Not Face or Ace}) = 1 - Pr(\text{Face or Ace}) = 1 - \dfrac{16}{52} = \dfrac{36}{52}.$$

5. Since there are $12 + 13 - 3 = 22$ face or club cards, $Pr(\text{Face or club}) = \dfrac{22}{52}$.

7.
$$\begin{array}{llllll} (1,1) & (1,2) & (1,3) & (1,4) & (1,5) & (1,6) \\ (2,1) & (2,2) & (2,3) & (2,4) & (2,5) & (2,6) \\ (3,1) & (3,2) & (3,3) & (3,4) & (3,5) & (3,6) \\ (4,1) & (4,2) & (4,3) & (4,4) & (4,5) & (4,6) \\ (5,1) & (5,2) & (5,3) & (5,4) & (5,5) & (5,6) \\ (6,1) & (6,2) & (6,3) & (6,4) & (6,5) & (6,6) \end{array}$$

All but 4 of the sums are neither greater than 5 nor odd. Hence:

$$Pr(\text{Odd or} >5) = 1 - \dfrac{4}{36} = \dfrac{32}{36}$$

9.
$$\begin{array}{llllll} (1,1) & (1,2) & (1,3) & (1,4) & (1,5) & (1,6) \\ (2,1) & (2,2) & (2,3) & (2,4) & (2,5) & (2,6) \\ (3,1) & (3,2) & (3,3) & (3,4) & (3,5) & (3,6) \\ (4,1) & (4,2) & (4,3) & (4,4) & (4,5) & (4,6) \\ (5,1) & (5,2) & (5,3) & (5,4) & (5,5) & (5,6) \\ (6,1) & (6,2) & (6,3) & (6,4) & (6,5) & (6,6) \end{array}$$

A consideration of the sums between 2 and 12 reveals the fact that only 5, 7, and 11 are not divisible by either 2 or 3. Noting that there are 12 such sums, we have:

$$Pr(\text{Divisible by 2 or 3}) = 1 - \dfrac{12}{36} = \dfrac{24}{36}$$

11. From the above sample space we see that there are 12 sums that are not divisible by either 2 or 3. Hence: $Pr(\text{Not by 2 or 3}) = \dfrac{12}{36}$.

13.
$$\begin{bmatrix} A & A & B & B & C & C \\ B & C & A & C & A & B \\ C & B & C & A & B & A \end{bmatrix}$$
$Pr(A \text{ in midle}) = \dfrac{2}{6}$

15.
$$\begin{bmatrix} A & A & B & B & C & C \\ B & C & A & C & A & B \\ C & B & C & A & B & A \end{bmatrix}$$
$Pr(A \text{ or B in middle}) = \dfrac{4}{6}$

17.
$$\begin{bmatrix} A & A & B & B & C & C \\ B & C & A & C & A & B \\ C & B & C & A & B & A \end{bmatrix}$$
$Pr(A \text{ on top or bottom}) = \dfrac{4}{6}$

19.
$$\begin{array}{c} R_1\ R_2\ R_3\ R_4 \\ W_1\ W_2\ W_3\ W_4\ W_5 \\ B_5\ B_{10} \end{array}$$
$Pr(R \text{ or even}) = \dfrac{4+2+1}{4+5+2} = \dfrac{7}{11}$

21. $\begin{Bmatrix} R_1\ R_2\ R_3\ R_4 \\ W_1\ \boxed{W_2}\ W_3\ \boxed{W_4}\ W_5 \\ B_5\ \boxed{B_{10}} \end{Bmatrix}$ $Pr(\text{not }R\text{ and even}) = \dfrac{2+1}{4+5+2} = \dfrac{3}{11}$

23. B_{10} is the only success. Hence:

$$Pr(B \text{ and } 10) = \frac{1}{11}$$

25. B_5 and B_{10} are the only successes. Hence: $Pr(B \text{ or } 10) = \dfrac{2}{11}$.

27. $\begin{Bmatrix} \boxed{\begin{matrix} R_1\ R_2\ R_3\ R_4 \\ W_1\ W_2\ W_3\ W_4\ W_5 \end{matrix}} \\ B_5\ B_{10} \end{Bmatrix}$ $Pr(\text{not }B\text{ and not }10) = \dfrac{4+5}{4+5+2} = \dfrac{9}{11}$, or: $1-\dfrac{2}{11} = \dfrac{9}{11}$.

29.

	1995	1996	1997	1998	1999	SUMS
Men Employed	60,085	64,897	66,524	67,134	67,761	326,401
Unemployed	3,239	3,147	2,826	2,580	2,433	14,225
Women Employed	54,396	53,310	57,647	57,278	58,655	281,286
Unemployed	2,819	2,783	2,187	2,424	2,285	12,498
SUMS	120,539	124,137	129,184	129,416	131,134	634,410

(a) $Pr(\text{Man}) = \dfrac{326{,}401+14{,}225}{634{,}410} = \dfrac{340{,}626}{634{,}410} \approx 0.54$

(b) $Pr(1996) = \dfrac{124{,}137}{634{,}410} \approx 0.20$

(c) $Pr(\text{Woman and } 1997) = \dfrac{57{,}647+2187}{634{,}410} = \dfrac{59{,}834}{634{,}410} \approx 0.094$

(d) $Pr(\text{Unemployed}) = \dfrac{14{,}225+12{,}498}{634{,}410} = \dfrac{26{,}723}{634{,}410} \approx 0.042$

(e) $Pr(\text{Not employed woman}) = 1 - Pr(\text{employed woman})$

$$= 1 - \frac{281{,}286}{634{,}410} = \frac{634{,}410-281{,}286}{634{,}410} = \frac{353{,}124}{634{,}410} \approx 0.557$$

(f) $Pr(\text{Unemployed and } 1999) = \dfrac{2433 + 2285}{634{,}410} = \dfrac{4718}{634{,}410} \approx 0.007$

(g) $Pr(\text{Woman and } 1999) = \dfrac{58{,}655 + 2285}{634{,}410} = \dfrac{60{,}940}{634{,}410} \approx 0.096$

31.

$525 + 50 + 5 + 5$ ⟶ 585

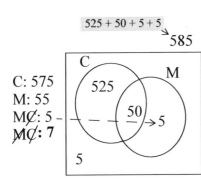

C: 575
M: 55
M$\cancel{C}$: 5
$\cancel{M}\cancel{C}$: 7

(a) $Pr(C \text{ or } M) = \dfrac{525 + 50 + 5}{85} = \dfrac{580}{585}$

(b) $Pr(\text{not } M) = \dfrac{525 + 5}{585} = \dfrac{530}{585}$

(c) $Pr(C \text{ not } C) = \dfrac{525}{585}$

(d) $Pr(M \text{ not } C) = \dfrac{5}{585}$

33.

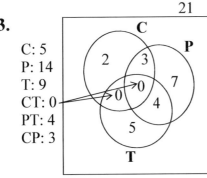

C: 5
P: 14
T: 9
CT: 0
PT: 4
CP: 3

(a) $2 + 7 + 5 = 14$ tutor only one course.

(b) None tutors all three courses (no one that tutors calculus tutors transitional).

(c) $Pr(\text{exactly two courses}) = \dfrac{3 + 4 + 0}{21} = \dfrac{7}{21}$

35.

D: 350
C: 275
B: 90
CD: 175
DB: 75
CB: 30
DCB: 25

(a) $95 + 5 + 10 + 290 = 400$ do not own a dog.

(b) $350 + 5 + 10 = 365$ own a dog or a bird.

(c) $750 - 290 = 460$ own at least one.

(d) $750 - 460 = 290$ own none.

37.

M: 23
MA: 10
A: 18
MH: 9
H: 14
AH: 8
MAH: 5

(a) $Pr(\text{only M}) = \dfrac{9}{110}$

(b) $Pr(\text{exactly two}) = \dfrac{4 + 5 + 3}{110} = \dfrac{12}{110}$

(c) $Pr(\text{none}) = \dfrac{77}{110}$

(d) $Pr(\text{exactly one}) = \dfrac{9 + 5 + 2}{110} = \dfrac{16}{110}$

(e) $Pr(\text{at most 1}) = \dfrac{77 + 9 + 5 + 2}{110} = \dfrac{93}{110}$

(f) $Pr(\text{at least 1}) = \dfrac{110 - 77}{110} = \dfrac{33}{110}$

39. Note: You don't want to have both a Female and a Male circle in your Venn diagram: the complement of the F-circle is the M-circle; just as the complement of the S-circle is the non-S circle.

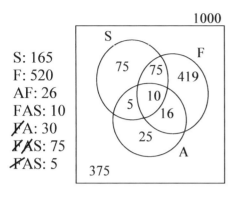

S: 165
F: 520
AF: 26
FAS: 10
F̸A: 30
F̸AS: 75
F̸AS: 5

(a) $Pr(S \text{ or } A) = \dfrac{165 + 25 + 16}{1000} = \dfrac{206}{1000}$

(b) $Pr(\text{not } S \text{ and not } A) = \dfrac{375 + 419}{1000} = \dfrac{794}{1000}$

(c) $Pr(A \text{ and not } S) = \dfrac{25 + 16}{1000} = \dfrac{41}{1000}$

(d) $Pr(S \text{ and } A) = \dfrac{5 + 10}{1000} = \dfrac{15}{1000}$

(e) $Pr(\text{not } F \text{ and } S) = \dfrac{75 + 5}{1000} = \dfrac{80}{1000}$

(f) $Pr(S \text{ or } A \text{ but not both}) = \dfrac{75 + 75 + 25 + 16}{1000} = \dfrac{191}{1000}$

(g) $Pr(\text{not } F \ \text{not } S \ \text{not } A) = \dfrac{375}{1000}$

(h) $Pr(F \ \text{not } S \ \text{not } A) = \dfrac{419}{1000}$

§1.3. Conditional Probability and Independent Events
Page 36

1. (a) Since there are 13 spades in the deck: $Pr(S) = \dfrac{13}{52}$.

(b) Since all 13 spades are black, and since there are 26 black cards: $Pr(S|B) = \dfrac{13}{26}$.

(c) Since there are still 13 spaces in the deck after the 13 diamonds are removed:
$$Pr(S| \text{ not } D) = \dfrac{13}{52 - 13} = \dfrac{13}{39}$$

3. (a) Since 4 of the 11 marbles are red: $Pr(R) = \dfrac{4}{11}$.

(b) Since your are given that the drawn marble is not white, there are $11 - 5 = 6$ possible marbles that can be drawn, 4 of which are red. Hence: $P(R|\text{not } W) = \dfrac{4}{6}$.

5. (a) $\begin{cases} (1,1)\ (1,2)\ (1,3)\ (1,4)\ (1,5)\ (1,6) \\ (2,1)\ (2,2)\ (2,3)\ (2,4)\ (2,5)\ (2,6) \\ (3,1)\ (3,2)\ (3,3)\ (3,4)\ (3,5)\ (3,6) \\ (4,1)\ (4,2)\ (4,3)\ (4,4)\ (4,5)\ (4,6) \\ (5,1)\ (5,2)\ (5,3)\ (5,4)\ (5,5)\ (5,6) \\ (6,1)\ (6,2)\ (6,3)\ (6,4)\ (6,5)\ (6,6) \end{cases}$
Half of the 26 rolls have an odd sum. So:
$Pr(\text{odd}) = \dfrac{18}{36}$

(b) $\begin{cases} (1,1)\ (1,2)\ (1,3)\ (1,4)\ (1,5)\ (1,6) \\ (2,1)\ (2,3)\ (2,5) \\ (3,1)\ (3,2)\ (3,3)\ (3,4)\ (3,5)\ (3,6) \\ (4,1)\ (4,3)\ (4,5) \\ (5,1)\ (5,2)\ (5,3)\ (5,4)\ (5,5)\ (5,6) \\ (6,1)\ (6,3)\ (6,5) \end{cases}$
There are $36 - 9 = 27$ possible outcomes of the experiment, with 3 successes in each of the 6 rows. So:
$$Pr(\text{odd sum}|\text{at least one odd}) = \dfrac{18}{27}$$

(c) The restriction that exactly one of the dice is odd reduces the sample space to 18 (each of the six rows contains three of them). They are all successes, since the sum of any odd number with an even number is odd. So: $Pr(\text{sum odd}|\text{exactly one die is odd}) = \dfrac{18}{18} = 1$.

7. Mark the forms with charitable deductions up to $500 with the letter A, those with deductions above $500 but not greater than $1000 with B, and those greater than $1000 with C. We are told that of the 500 forms 50 are A's, 250 are B's, and 200 are C's (including Mr. G). So:

(a) $Pr(\text{Mr. G}) = \dfrac{1}{500}$ (b) $Pr(\text{Mr. G}|B \text{ or } C) = \dfrac{1}{250 + 200} = \dfrac{1}{450}$ (c) $Pr(\text{Mr. G}|C) = \dfrac{1}{200}$

9.

		District A	District B	District C	District D	Sum
MALE						
	Democrat	2500	1863	1952	2430	8745
	Republican	1400	2400	2100	1300	7200
	Independent	750	920	1011	951	3632
FEMALE						
	Democrat	2620	2001	1973	2456	9050
	Republican	1100	2122	2351	1342	6915
	Independent	624	830	1121	1012	3587
	Sum	8994	10,136	10,508	9491	**39,129**

(a) There are $2500 + 2620 = 5120$ Democrats in District A which lists a total of 8994 individuals. So: $Pr(\text{Democrat}|\text{District A}) = \dfrac{5120}{8994} \approx 0.57$.

(b) There are a total of $8745 + 9059 = 17,804$ Democrats, with $2500 + 2620 = 5120$ of them in District A. So: $Pr(\text{District A}|\text{Democrat}) = \dfrac{5120}{17,804} \approx 0.29$.

(c) In the group of 17,796 Democrats, 8745 are males. So: $Pr(\text{male}|\text{Democrat}) = \dfrac{8745}{17,795} \approx 0.$:

(d) The sum all the individuals in Districts A, B, and D is $8994 + 10,136 + 9491 = 28,621$, of which $2500 + 1400 + 750 + 1863 + 2400 + 920 + 2430 + 1300 + 951 = 14,751$ are males. So: $Pr(\text{male}|\text{not in District } C) = \dfrac{14,751}{28,621} \approx 0.52$.

(e) The sum of all the individuals in Districts A or C that are Democrats or Republicans is $8994 - (750 + 624) + 10,508 - (1011 + 1121) = 15,996$, of which $2620 + 1100 + 1973 + 2351 = 8044$ are female. So: $Pr(\text{female}|\text{in A or C}) = \dfrac{8044}{15,996} \approx 0.50$.

(f) There are $39,129 - (2500 + 1400 + 750) - (1973 + 2351 + 1121) = 29,034$ individuals that are not males in A and not females in C. Of which $(8745 - 2500) + (7200 - 1400) + (9050 - 1973) + (6915 - 2351) = 23,686$ are Democrats or Republicans. So: $Pr(\text{not Ind}|\text{not M in A and not F in C}) = \dfrac{23,686}{29,034} \approx 0.82$.

11. (a) $Pr(RRR) = Pr(R_{first})Pr(R_{second}|R_{first})Pr(R_{third}|\text{first 2 were }R) = \frac{4}{11}\cdot\frac{3}{10}\cdot\frac{2}{9} = \frac{4}{165} \approx 0.024$

(b) $Pr(\cancel{R}\cancel{R}\cancel{R}) = Pr(\cancel{R}_{first})Pr(\cancel{R}_{second}|\cancel{R}_{first})Pr(\cancel{R}_{third}|\text{first 2 were not }R) = \frac{7}{11}\cdot\frac{6}{10}\cdot\frac{5}{9} = \frac{7}{33} \approx 0.21$

(c) $Pr(R_{first}\cancel{R}\cancel{R}) = Pr(R_{first})Pr(\cancel{R}_{second}|R_{first})Pr(\cancel{R}_{third}|R\text{ and }\cancel{R}\text{ gone}) = \frac{4}{11}\cdot\frac{7}{10}\cdot\frac{6}{9} = \frac{28}{165} \approx 0.17$

(d) $Pr(R_{first}\cancel{R}\cancel{R}) + Pr(\cancel{R}R_{second}\cancel{R}) + Pr(\cancel{R}\cancel{R}R_{third}) = 3\left(\frac{28}{165}\right) = \frac{28}{55} \approx 0.51$

see (c)

13. $Pr(DDD) = Pr(D_{first})Pr(D_{second}|D_{first})Pr(D_{third}|\text{first 2 were }D) = \frac{13}{52}\cdot\frac{12}{51}\cdot\frac{11}{50} = \frac{11}{850} \approx 0.013$

15. $Pr(DDD) + Pr(CCC) + Pr(HHH) + Pr(SSS) = 4\left(\frac{11}{850}\right) = \frac{22}{425} \approx 0.052$

see 13

17. $Pr(A_{first}A_{second}\cancel{A}) = Pr(A_{first})Pr(A_{second}|A_{first})Pr(\cancel{A}_{third}|\text{two }A\text{ gone}) = \frac{4}{52}\cdot\frac{3}{51}\cdot\frac{48}{50} \approx 0.0043$

19. $Pr(CCC|F) = Pr(C_{first})Pr(C_{second}|C_{first})Pr(C_{third}|\text{first 2 were }C) = \frac{3}{12}\cdot\frac{2}{11}\cdot\frac{1}{10} \approx 0.0045$

Restricts the sample space to the 12 face cards, which contains 3 clubs

21. Assume, without loss of generality, that they are all Hearts. Then:

$Pr(FFF|H) = Pr(F_{first})Pr(F_{second}|F_{first})Pr(F_{third}|\text{first 2 were }F) = \frac{3}{13}\cdot\frac{2}{12}\cdot\frac{1}{11} \approx 0.0035$

Restrict the sample space to the 13 hearts

23. $Pr(FFF|\cancel{X}) = \frac{8}{48}\cdot\frac{7}{47}\cdot\frac{6}{46} = \frac{7}{2162} \approx 0.0032$

Restricts the sample space to the 48 cards, which contains 8 face cards.

25. $Pr(A_{first}K_{second}Q_{third}|\cancel{X}) = \frac{4}{48}\cdot\frac{4}{47}\cdot\frac{4}{46} = \frac{2}{3243} \approx 0.00062$

Restricts the sample space to the 48 cards, which contains all 4 of the Aces, Kings, and Queens

27. (a) $Pr(R_{first}) = \dfrac{1}{6}$ (b) $Pr(R_{first}|G) = \dfrac{1}{5}$ (c) $Pr(R_{last}|G) = \dfrac{1}{5}$

restricted sample space   restricted sample space

(d) $Pr(R_{last}|G_{first}) = \dfrac{1}{5}$ Since it is given that the Green is hoisted first, there are 5 flags that can be hoisted second, one of which is Red,

29. (a) Since these are five independent events, and since the probability of drawing an Ace is $\dfrac{4}{52}$, the probability that all will draw an Ace is $\left(\dfrac{4}{52}\right)^5 \approx 0.000003$.

(b) Since these are five independent events, and since the probability of drawing an Ace of a King is $\dfrac{8}{52}$, the probability that all will draw an Ace or a King is $\left(\dfrac{8}{52}\right)^5 \approx 0.000086$.

31. (a) Since these are five independent events, and since the probability of picking a 7 is $\dfrac{1}{10}$, the probability that all will pick a 7 is $\left(\dfrac{1}{10}\right)^5 = 0.00001$.

(b) The probability that all will pick the same number is 10 times the probability that all will pick, any of the 10 numbers. So: $Pr(\text{all same}) = 10\left(\dfrac{1}{10}\right)^5 = 0.0001$ [see (a)].

33. (a) If we conveniently assume that the probability of what happened when Every faces the 356th batter will not alter the probability of what will happen when he faces the next batter, then:
$$Pr(\text{retire the next two batters}) = \left(\dfrac{355-75}{355}\right)^2 = \left(\dfrac{280}{355}\right)^2 \approx 0.6221$$

(b) Operating under the assumption of part (a):
$$Pr(\text{retire neither of the next two batters}) = \left(\dfrac{75}{355}\right)^2 = 0.0446$$

(c) $Pr(\text{retires the next two batters}) = \dfrac{280}{355} \cdot \dfrac{281}{356} \approx 0.6226$ not much different from the 0.6221 of part (a). (He had already faced 355 batters.)

since he retired the 356th batter, he then has retired 281 of the 356 batters faced.

$Pr(\text{retires neither ofthe next two batters}) = \dfrac{75}{355} \cdot \dfrac{76}{356} \approx 0.0451$ not much different from the 0.0446 of part (b)

since he retired the 356th batter, he retired 281 of the 356 batters faced

35. (a) $Pr(\cancel{5}\,\cancel{5}\,5) = \dfrac{5}{6} \cdot \dfrac{5}{6} \cdot \dfrac{1}{6} \approx 0.12$

(b) $Pr(5) + Pr(\cancel{5}\,5) + Pr(\cancel{5}\,\cancel{5}\,5) = \dfrac{1}{6} + \dfrac{5}{6} \cdot \dfrac{1}{6} + \dfrac{5}{6} \cdot \dfrac{5}{6} \cdot \dfrac{1}{6} \approx 0.42$

At least 3 times is **not** 1 or 2

(c) $Pr(\text{at least three}) = 1 - [Pr(5) + Pr(\cancel{5}\,5)] = 1 - \left(\dfrac{1}{6} + \dfrac{5}{6} \cdot \dfrac{1}{6}\right) \approx 0.69$

37. (a) $Pr(A) = \dfrac{4}{52} \approx 0.077$ (b) $Pr(\cancel{A}A) = \dfrac{48}{52} \cdot \dfrac{4}{51} \approx 0.072$

(c) $Pr(\cancel{A}\cancel{A}A) = \dfrac{48}{52} \cdot \dfrac{47}{51} \cdot \dfrac{4}{50} \approx 0.068$ (d) $Pr(\cancel{A}\cancel{A}\cancel{A}A) = \dfrac{48}{52} \cdot \dfrac{47}{51} \cdot \dfrac{46}{50} \cdot \dfrac{4}{49} \approx 0.064$

(e) The worst that can happen is that you keep drawing cards till only the four aces remain (that will take 48 draws). The game must then end on the next or 49th draw.

39. (a) $Pr(F\cancel{F}\cancel{F}) = \dfrac{40}{52} \cdot \dfrac{39}{51} \cdot \dfrac{12}{50} \approx 0.141$

(b) $Pr(\text{at most } 3) = Pr(F) + Pr(\cancel{F}F) + Pr(\cancel{F}\cancel{F}F) = \dfrac{12}{52} + \dfrac{40}{52} \cdot \dfrac{12}{51} + \dfrac{40}{52} \cdot \dfrac{39}{51} \cdot \dfrac{12}{50} \approx 0.553$

At least 3 times is **not** 1 or 2
$\downarrow$

(c) $Pr(\text{at least three}) = \mathbf{1} - [Pr(F) + Pr(F\cancel{F})] = 1 - \left(\dfrac{12}{52} + \dfrac{40}{52} \cdot \dfrac{12}{51}\right) \approx 0.588$

41. (a) Since the better player hallway wins, the best player must get the golden gidget (probability equals 1).

(b) In order for the second best player to come in second, he cannot come against the best player until the last round of the tournament. So, of the 15 possible opponents in the first round, he must not draw the name of the best player (14 of the 15 will do). In the second round he can beat 6 of the 7 players; and in the third round he can beat 2 of the 3 players. So:

$$Pr(\text{second best player wins the silver gidget}) = \dfrac{14}{15} \cdot \dfrac{6}{7} \cdot \dfrac{2}{3} \approx 0.53$$

43. (a) Not to malfunctions requires that A does not fail (prob: $\dfrac{9}{10}$) **and** that **NOT** both B and C fail

[prob: $1 - (\dfrac{2}{15} \cdot \dfrac{2}{9})$]. So $Pr(\text{not malfunctions}) = \dfrac{9}{10}[1 - (\dfrac{2}{15} \cdot \dfrac{2}{9})] = \dfrac{131}{150} \approx 0.87$.

(b) $Pr(\text{malfunctions}) = 1 - Pr(\text{not malfunctions}) = 1 - \dfrac{131}{150} = \dfrac{19}{150} \approx 0.13$.

§1.4. Trees and Bayes' Formula
Page 50

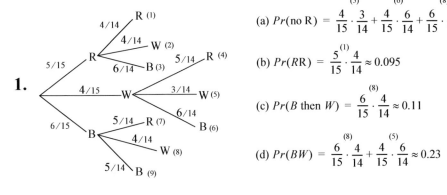

1.

(a) $Pr(\text{no R}) = \overset{(5)}{\frac{4}{15}} \cdot \frac{3}{14} + \overset{(6)}{\frac{4}{15}} \cdot \frac{6}{14} + \overset{(8)}{\frac{6}{15}} \cdot \frac{4}{14} + \overset{(9)}{\frac{6}{15}} \cdot \frac{5}{14} = 0.43$

(b) $Pr(RR) = \overset{(1)}{\frac{5}{15}} \cdot \frac{4}{14} \approx 0.095$

(c) $Pr(B \text{ then } W) = \frac{6}{15} \cdot \overset{(8)}{\frac{4}{14}} \approx 0.11$

(d) $Pr(BW) = \frac{6}{15} \cdot \overset{(8)}{\frac{4}{14}} + \frac{4}{15} \cdot \overset{(5)}{\frac{6}{14}} \approx 0.23$

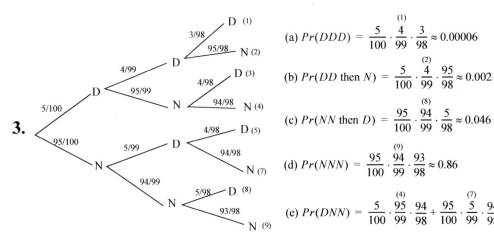

3.

(a) $Pr(DDD) = \frac{5}{100} \cdot \frac{4}{99} \cdot \overset{(1)}{\frac{3}{98}} \approx 0.00006$

(b) $Pr(DD \text{ then } N) = \frac{5}{100} \cdot \frac{4}{99} \cdot \overset{(2)}{\frac{95}{98}} \approx 0.002$

(c) $Pr(NN \text{ then } D) = \frac{95}{100} \cdot \frac{94}{99} \cdot \overset{(8)}{\frac{5}{98}} \approx 0.046$

(d) $Pr(NNN) = \frac{95}{100} \cdot \frac{94}{99} \cdot \overset{(9)}{\frac{93}{98}} \approx 0.86$

(e) $Pr(DNN) = \frac{5}{100} \cdot \frac{95}{99} \cdot \overset{(4)}{\frac{94}{98}} + \frac{95}{100} \cdot \frac{5}{99} \cdot \overset{(7)}{\frac{94}{98}} + \frac{95}{100} \cdot \frac{94}{99} \cdot \overset{(8)}{\frac{5}{98}} \approx 0.046$

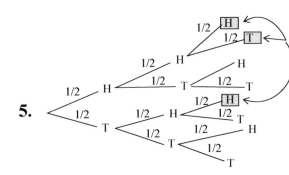

The boxed-in "leaves" are the successes. So:

$Pr(2 \text{ or } 3 \text{ consecutuv heads}) = \left(\frac{1}{2}\right)^3 + \left(\frac{1}{2}\right)^3 + \left(\frac{1}{2}\right)^3 = \frac{3}{8}$

5.

7.

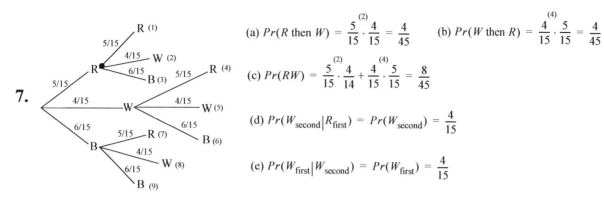

(a) $Pr(R \text{ then } W) = \overset{(2)}{\dfrac{5}{15}} \cdot \dfrac{4}{15} = \dfrac{4}{45}$ (b) $Pr(W \text{ then } R) = \dfrac{4}{15} \cdot \overset{(4)}{\dfrac{5}{15}} = \dfrac{4}{45}$

(c) $Pr(RW) = \overset{(2)}{\dfrac{5}{15}} \cdot \dfrac{4}{14} + \dfrac{4}{15} \cdot \overset{(4)}{\dfrac{5}{15}} = \dfrac{8}{45}$

(d) $Pr(W_{\text{second}} | R_{\text{first}}) = Pr(W_{\text{second}}) = \dfrac{4}{15}$

(e) $Pr(W_{\text{first}} | W_{\text{second}}) = Pr(W_{\text{first}}) = \dfrac{4}{15}$

9.

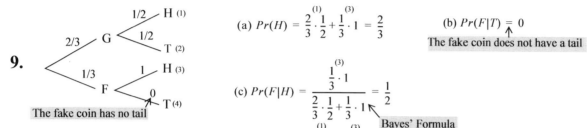

The fake coin has no tail

(a) $Pr(H) = \overset{(1)}{\dfrac{2}{3}} \cdot \dfrac{1}{2} + \overset{(3)}{\dfrac{1}{3}} \cdot 1 = \dfrac{2}{3}$ (b) $Pr(F|T) = 0$
    The fake coin does not have a tail

(c) $Pr(F|H) = \dfrac{\overset{(3)}{\dfrac{1}{3}} \cdot 1}{\underset{(1)}{\dfrac{2}{3}} \cdot \dfrac{1}{2} + \underset{(3)}{\dfrac{1}{3}} \cdot 1} = \dfrac{1}{2}$
    Bayes' Formula

11.

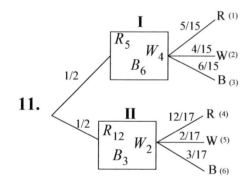

(a) $Pr(R) = \overset{(1)}{\dfrac{1}{2}} \cdot \dfrac{5}{15} + \dfrac{1}{2} \cdot \overset{(4)}{\dfrac{12}{17}} \approx 0.52$ (b) $Pr(R|II) = \dfrac{12}{17} \approx 0.71$
    You are given that the marble is drawn from urn 2.

(c) $Pr(II|R) = \dfrac{\dfrac{1}{2} \cdot \overset{(4)}{\dfrac{12}{17}}}{\underset{(1)}{\dfrac{1}{2}} \cdot \dfrac{5}{15} + \dfrac{1}{2} \cdot \underset{(4)}{\dfrac{12}{17}}} \approx 0.68$

Bayes' Formula

(d) $Pr(I|R) = \dfrac{\dfrac{1}{2} \cdot \overset{(5)}{\dfrac{4}{15}} + \dfrac{1}{2} \cdot \overset{(6)}{\dfrac{6}{15}}}{1 - \left(\underset{(1)}{\dfrac{1}{2}} \cdot \dfrac{5}{15} + \dfrac{1}{2} \cdot \underset{(4)}{\dfrac{12}{17}} \right)} \approx 0.70$

13.

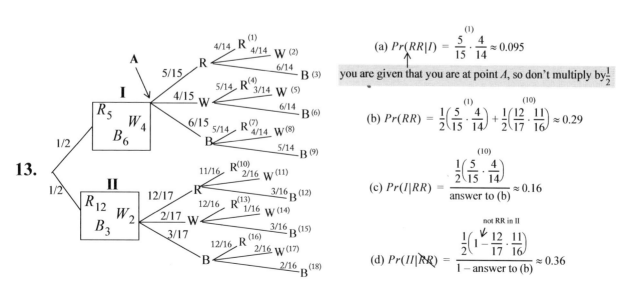

(a) $Pr(RR|I) = \overset{(1)}{\dfrac{5}{15}} \cdot \dfrac{4}{14} \approx 0.095$
    you are given that you are at point A, so don't multiply by $\dfrac{1}{2}$

(b) $Pr(RR) = \dfrac{1}{2}\left(\overset{(1)}{\dfrac{5}{15}} \cdot \dfrac{4}{14} \right) + \dfrac{1}{2}\left(\dfrac{12}{17} \cdot \overset{(10)}{\dfrac{11}{16}} \right) \approx 0.29$

(c) $Pr(I|RR) = \dfrac{\dfrac{1}{2}\left(\dfrac{5}{15} \cdot \overset{(10)}{\dfrac{4}{14}} \right)}{\text{answer to (b)}} \approx 0.16$

(d) $Pr(II|RR) = \dfrac{\dfrac{1}{2}\left(1 - \overset{\text{not RR in II}}{\dfrac{12}{17} \cdot \dfrac{11}{16}} \right)}{1 - \text{answer to (b)}} \approx 0.36$

15.

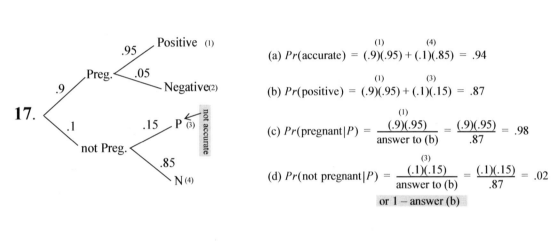

10/60 P (1)

20/60 B (2)

F   15/60 A (3)

15/60 U (4)

60/100

40/100   12/40 P (5)

4/40 B (6)

M   7/40 C (7)

5/40

12/40 E (8)

U (9)

(a) $Pr(B) = \overset{(2)}{\dfrac{60}{100}} \cdot \dfrac{20}{60} + \overset{(6)}{\dfrac{40}{100}} \cdot \dfrac{4}{40} = \dfrac{6}{25}$ (b) $Pr(F \text{ and } B) = \overset{(2)}{\dfrac{60}{100}} \cdot \dfrac{20}{60} = \dfrac{1}{5}$

(c) $Pr(E) = \dfrac{40}{100} \cdot \overset{(8)}{\dfrac{5}{40}} = \dfrac{1}{20}$

(d) $Pr(F|B) = \dfrac{\text{answer to (b)}}{\text{answer to (a)}} = \dfrac{1/5}{6/25} = \dfrac{5}{6}$ $\boxed{\text{Bayes' Formula}}$

(e) $Pr(B|F) = \dfrac{20}{60} = \dfrac{1}{3}$ we are given that the student if a female

(d) $Pr(F|B) = 1$ every art major is a female

17.

.95 Positive (1)

Preg. .05

.9   Negative (2)

.1

not Preg.   .15 P (3) not accurate

.85

N (4)

(a) $Pr(\text{accurate}) = \overset{(1)}{(.9)(.95)} + \overset{(4)}{(.1)(.85)} = .94$

(b) $Pr(\text{positive}) = \overset{(1)}{(.9)(.95)} + \overset{(3)}{(.1)(.15)} = .87$

(c) $Pr(\text{pregnant}|P) = \dfrac{\overset{(1)}{(.9)(.95)}}{\text{answer to (b)}} = \dfrac{(.9)(.95)}{.87} = .98$

(d) $Pr(\text{not pregnant}|P) = \dfrac{\overset{(3)}{(.1)(.15)}}{\text{answer to (b)}} = \dfrac{(.1)(.15)}{.87} = .02$

or 1 − answer (b)

19.

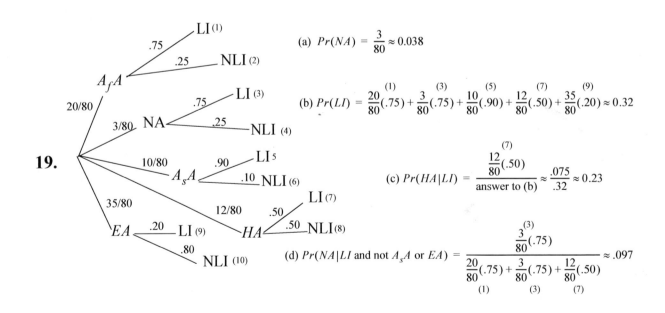

.75 LI (1)

.25 NLI (2)

A_fA

20/80

3/80 NA   .75 LI (3)

.25 NLI (4)

10/80   .90 LI 5

A_sA .10 NLI (6)

35/80   12/80 .50 LI (7)

HA   .50 NLI (8)

EA   .20 LI (9)

.80

NLI (10)

(a) $Pr(NA) = \dfrac{3}{80} \approx 0.038$

(b) $Pr(LI) = \overset{(1)}{\dfrac{20}{80}(.75)} + \overset{(3)}{\dfrac{3}{80}(.75)} + \overset{(5)}{\dfrac{10}{80}(.90)} + \overset{(7)}{\dfrac{12}{80}(.50)} + \overset{(9)}{\dfrac{35}{80}(.20)} \approx 0.32$

(c) $Pr(HA|LI) = \dfrac{\overset{(7)}{\dfrac{12}{80}(.50)}}{\text{answer to (b)}} \approx \dfrac{.075}{.32} \approx 0.23$

(d) $Pr(NA|LI \text{ and not } A_sA \text{ or } EA) = \dfrac{\overset{(3)}{\dfrac{3}{80}(.75)}}{\underset{(1)}{\dfrac{20}{80}(.75)} + \underset{(3)}{\dfrac{3}{80}(.75)} + \underset{(7)}{\dfrac{12}{80}(.50)}} \approx .097$

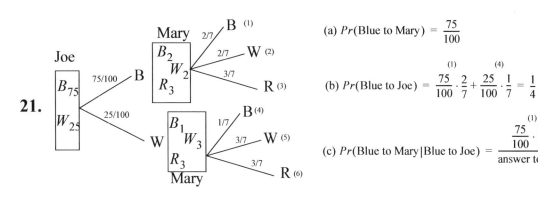

21.

(a) $Pr(\text{Blue to Mary}) = \dfrac{75}{100}$

(b) $Pr(\text{Blue to Joe}) = \overset{(1)}{\dfrac{75}{100}} \cdot \dfrac{2}{7} + \overset{(4)}{\dfrac{25}{100}} \cdot \dfrac{1}{7} = \dfrac{1}{4}$

(c) $Pr(\text{Blue to Mary}|\text{Blue to Joe}) = \dfrac{\overset{(1)}{\dfrac{75}{100} \cdot \dfrac{2}{7}}}{\text{answer to (b)}} \approx \dfrac{.21}{.25} \approx 0.84$

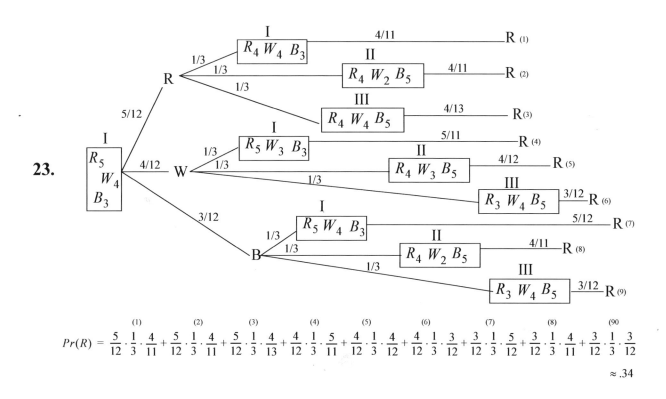

23.

$$Pr(R) = \overset{(1)}{\dfrac{5}{12} \cdot \dfrac{1}{3} \cdot \dfrac{4}{11}} + \overset{(2)}{\dfrac{5}{12} \cdot \dfrac{1}{3} \cdot \dfrac{4}{11}} + \overset{(3)}{\dfrac{5}{12} \cdot \dfrac{1}{3} \cdot \dfrac{4}{13}} + \overset{(4)}{\dfrac{4}{12} \cdot \dfrac{1}{3} \cdot \dfrac{5}{11}} + \overset{(5)}{\dfrac{4}{12} \cdot \dfrac{1}{3} \cdot \dfrac{4}{12}} + \overset{(6)}{\dfrac{4}{12} \cdot \dfrac{1}{3} \cdot \dfrac{3}{12}} + \overset{(7)}{\dfrac{3}{12} \cdot \dfrac{1}{3} \cdot \dfrac{5}{12}} + \overset{(8)}{\dfrac{3}{12} \cdot \dfrac{1}{3} \cdot \dfrac{4}{11}} + \overset{(90)}{\dfrac{3}{12} \cdot \dfrac{1}{3} \cdot \dfrac{3}{12}}$$

$$\approx .34$$

---

§1.5 The Fundamental Counting Principle
Page 60

1.
$$\underset{26}{\dfrac{X}{}} \quad \underset{25}{\dfrac{Y}{}} \quad \underset{1}{\dfrac{3}{}} \quad \underset{1}{\dfrac{3}{}} \quad \underset{1}{\dfrac{3}{}} \quad \underset{1}{\dfrac{3}{}}$$
$\uparrow$——— choices ———$\uparrow$

Letters are different and the numbers are the same:

$$26 \cdot 25 \cdot 10 \cdot 1 \cdot 1 \cdot 1 = 6500$$

3.
$$\underset{1}{\dfrac{A}{}} \quad \underset{26}{\dfrac{Y}{}} \quad \underset{10}{\dfrac{3}{}} \quad \underset{10}{\dfrac{4}{}} \quad \underset{10}{\dfrac{3}{}} \quad \underset{10}{\dfrac{2}{}}$$
$\uparrow$——— choices ———$\uparrow$

Letters A first (no other restrictions):

$$1 \cdot 26 \cdot 10 \cdot 10 \cdot 10 \cdot 10 = 260,000$$

5. $\dfrac{A}{1}\ \dfrac{Y}{25}\ \dfrac{3}{10}\ \dfrac{4}{10}\ \dfrac{3}{10}\ \dfrac{2}{10}$ or $\dfrac{Z}{25}\ \dfrac{A}{1}\ \dfrac{0}{10}\ \dfrac{9}{10}\ \dfrac{3}{10}\ \dfrac{2}{10}$ A appears exactly once:

$\uparrow$————— choices ————$\uparrow$ $\uparrow$———— choices ————$\uparrow$

$$1 \cdot 25 \cdot 10 \cdot 10 \cdot 10 \cdot 10 + 25 \cdot 1 \cdot 10 \cdot 10 \cdot 10 \cdot 10 = 500{,}000 \leftarrow$$

7. $\dfrac{C}{26}\ \dfrac{A}{25}\ \dfrac{6}{4}\ \dfrac{7}{4}\ \dfrac{7}{4}\ \dfrac{9}{4}$ there are four digits greater than 5: 6, 7, 8, and 9

$\uparrow$———— choices ————$\uparrow$

Letters different and digits greater than 5: $26 \cdot 25 \cdot 4 \cdot 4 \cdot 4 \cdot 4 = 166{,}400$

9. $\dfrac{X}{26}\ \dfrac{X}{1}\ \dfrac{0}{10}\ \dfrac{4}{10}\ \dfrac{3}{10}\ \dfrac{0}{1}$ no restriction of the first three digits

$\uparrow$————— choices ————$\uparrow$

Letters same and last digit is 0: $26 \cdot 1 \cdot 10 \cdot 10 \cdot 10 \cdot 1 = 26{,}000$

11. Think of this as being a two-step process: Step 1 the two letter possibilities ($26 \cdot 26$ choices), and Step 2: the four digit possibilities (namely that the digit increase and that the last digit is 4). How many choices do we have in that second step? Let's just list and count them:

$$1\,2\,3\,4 — 0\,1\,2\,4 — 0\,1\,3\,4 — 0\,2\,3\,4 \text{ (a total of 4 choices).}$$

Conclusion There are $26 \cdot 26 \cdot 4 = 2704$ plates with increasing digits ending in a 4.

13. $\boxed{\underline{\quad \text{prefix} \quad}}$ no restriction for these 7 digits
choice o 2 $\dfrac{}{10}\ \dfrac{}{10}\ \dfrac{}{10}\ \dfrac{}{10}\ \dfrac{}{10}\ \dfrac{}{10}\ \dfrac{}{10}$ Possible number: 2×10^7

15. $\boxed{\dfrac{5}{\ }\ \dfrac{7}{\ }\ \dfrac{7}{\ }}$ no restriction for these 7 digits
$\uparrow$ $\dfrac{}{10}\ \dfrac{}{10}\ \dfrac{}{10}\ \dfrac{}{10}\ \dfrac{}{10}\ \dfrac{}{10}\ \dfrac{}{10}$ Possible number: $3 \cdot 9 \times 10^7$
$\uparrow$

choice of 3 of where **not** to put the 7 followed by a **choice of 9** in that position (we chose "5" but could have chosen any digit distinct from 7). A choice of 1 for the remaining two digits in the prefix (the remaining two digits must be 7's)

3 digits are divisible by 4: 0, 4, and 8

17. $\dfrac{3}{5}\ \dfrac{7}{5}\ \dfrac{1}{5}\ \dfrac{4}{3}\ \dfrac{4}{3}\ \dfrac{0}{3}\ \dfrac{0}{3}\ \dfrac{8}{3}\ \dfrac{4}{3}\ \dfrac{8}{3}$ Possible number: $5^3 \cdot 3^7 = 273{,}375$

$\uparrow$— choices ————$\uparrow$

19. Choice of 2 soups, followed by a choice of 3 salads, followed by a choice of 4 dressings, followed by a choice of 9 entrées, followed by a choice of 5 desserts.

Choices followed by choices, you multiply: $\mathbf{2 \cdot 3 \cdot 4 \cdot 9 \cdot 5} = 1080$

21. If there are no imposed restrictions, then there are 10^9 possible (9 digit) social security numbers.

(a) $Pr(\text{first three digits are the same}) = \dfrac{\overset{\text{choice of 10 for first digit}}{10} \cdot \overset{\text{next two digits must be the same as the first}}{1 \cdot 1} \cdot \overset{\text{no restriction on the last 6 digits}}{10^6}}{10^9} = \dfrac{10^7}{10^9} = \dfrac{1}{100}$.

(b)

5	1	3	5	4	3	3	3	3
9	9	9	9	9	9	1	1	1

choices

can't be 0: choice if 9   must be the same: choice of 1

$Pr(\text{no 0 and last 4 the same}) = \dfrac{9^6 \cdot 1 \cdot 1 \cdot 1}{10^9} \approx .00053$

(c)

choice of 9 (can't be 0)   not 0 and all different

5	1	3	5	4	3	2	4	7
9	9	9	9	9	9	8	7	6

$Pr(\text{no 0 and last 4 different}) = \dfrac{9^6 \cdot 8 \cdot 7 \cdot 6}{10^9} \approx 0.18$

(d)

last digit must be 0

5	1	3	5	4	3	2	4	0
9	8	7	6	5	4	3	2	1

cant be 0 cant be 0 or the first digit; etc.

$Pr(\text{no repetition and last digit is 0})$
$= \dfrac{9 \cdot 8 \cdot 7 \cdot 6 \cdot 5 \cdot 4 \cdot 3 \cdot 2 \cdot 1}{10^9} \approx 0.00036$

23. [For (a), (b), and (c): There are 4^4 possible 4-link molecules (4 choices four times).]

(a) C C C C ← a success

4	1	1	1

choice of 4   then no more choices

$Pr(\text{contains but one sub-molecule}) = \dfrac{4 \cdot 1 \cdot 1 \cdot 1}{4^4} = \dfrac{1}{4^3} \approx 0.016$

(b) C A T G ← a success

  4   3 2 1

no repetition

$Pr(\text{no repetition}) = \dfrac{4 \cdot 3 \cdot 2 \cdot 1}{4^4} = \dfrac{24}{4^4} \approx 0.093$

(c)

successes

A A T A   or   G G G G Choice of 4: which letter is to appear each time (we chose G)

Choose the letter (**choice of 4**: we chose A).
Chose where the chosen letter does not appear (**choice of 4**: we chose the third spot).
Chose the other letter (**choice of 3**: we chose T).
Total number of choices when a sub-molecule appears exactly 3 times: $4 \cdot 4 \cdot 3$.

$Pr(\text{One sub-molecule 3 or 4 times}) = \dfrac{4 \cdot 4 \cdot 3}{4^4} + \dfrac{4}{4^4} \approx 0.20$

Note: Here is a wrong approach: Chose the letter (choice of 4); put it in three places (Choice of 4); put any letter (including the chosen letter) in the remaining spot (if it is not the chosen letter, then the chosen letter will appear exactly 3 times; if it is the chosen letter, then it will appear four times). Why is it wrong? A tricky question. The problem is that some outcomes will be counted more than once.

[For (d) and (e): There are 4^3 possible 3-link molecules (4 choices three times).]

(d) C A A ← a success

Chose where A does not sit (**choice of 3**: we chose the first spot)
Chose which letter sits in that spot (**choice of 3**: we chose C)

$$Pr(\text{A exactly twice}) = \frac{3 \cdot 3}{4^3} \approx 0.14$$

(e)

C A A   **or**   T T T

Chose the letter (choice of 4: we chose A)
Chose where the chosen letter does not appear (choice of 3)
Chose the other letter (choice of 3: we chose C)

Choose the letter

$$Pr(\text{One sub-molecule 2 or 3 times}) = \frac{4 \cdot 3 \cdot 3}{4^3} + \frac{4}{4^3} = \frac{5}{8} \approx 0.63$$

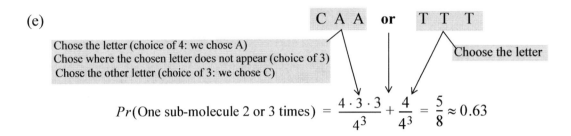

25. (a) 7 7 7 ← a success
          9 1 1
          ↑
       can't be 0

$$Pr(\text{all same}) = \frac{9 \cdot 1 \cdot 1}{9 \cdot 10 \cdot 10} = \frac{1}{100}$$

↑
choice of 9 (can't be zero), followed by any two other digits.

(b) 7 8 3 ← a success
     9 10 5 ← must be odd
     ↑
  can't be 0

$$Pr(\text{odd}) = \frac{9 \cdot 10 \cdot 5}{9 \cdot 10 \cdot 10} = \frac{1}{2}$$

(c) 1 5 3 ← a success
     9 9 9
       ↖↑↗
      no 0

$$Pr(\text{no 0}) = \frac{9 \cdot 9 \cdot 9}{9 \cdot 10 \cdot 10} = \frac{81}{100}$$

(d)
successes

1 0 3   **or**   1 3 0   **or**   1 0 0
9 1 9   9 9 1   9 1 1

put a zero in   put a zero at   two zeros
the middle only   end only

$$Pr(\text{0 appears}) = \frac{2(9 \cdot 1 \cdot 9) + (9 \cdot 1 \cdot 1)}{9 \cdot 10 \cdot 10} = \frac{19}{100}$$

(e)
successes

1 0 3   **or**   1 3 0
9 1 9   9 9 1

put a zero in   put a zero at
the middle only   end only

$$Pr(\text{0 appears exactly once}) = \frac{2(9 \cdot 1 \cdot 9)}{9 \cdot 10 \cdot 10} = \frac{18}{100}$$

27. If the table were not round, there would be $6 \cdot 5 \cdot 4 \cdot 3 \cdot 2 \cdot 1$ different ways of seating the 6 individual [See (a) below].

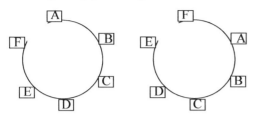
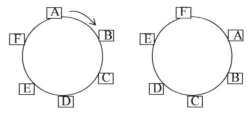

The above are two different seating arrangements since in the left F is not seated next to A, while F is seated next to A on the right.

The above are the same seating arrangements in that the neighbors of every individual is the same in both. Indeed, you can rotate the one on the left 5 times and still end up with the same seating arrangement.

So: Every seating arrangement at the round table gives rise to 6 different seating arrangements at a straight table.

Conclusion: There are $\dfrac{6 \cdot 5 \cdot 4 \cdot 3 \cdot 2 \cdot 1}{6}$ different seating arrangements of six individuals at a round table.

(a)

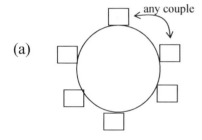

Take couple one and seat it anywhere (Since the table is round, we can assume that they occupy the position in the adjacent figure). We do have **2 choices** however: Male and then Female or the other way around. Moving in a clockwise direction we **chose one of the 2** remaining couples and sit them down, with a **choice of 2** (M and F or F and M). No choice as to what couple is seated next, but there is again a seating **choice of 2**: M and F or F and M.

$$Pr(\text{M and F of each couple together}) = \frac{2 \cdot 2 \cdot 2 \cdot 2}{5 \cdot 4 \cdot 3 \cdot 2 \cdot 1} = \frac{2}{15}$$

(b) The particular chair which seats the first male (thinking "clockwise") is irrelevant, but there are $3 \cdot 2 \cdot 1$ **choices** of seating the three males in those chairs. The three females have to sit in the remaining three chairs, but again there are $3 \cdot 2 \cdot 1$ **choices** of seating them in those chairs.

So: $Pr(\text{males seated nex to each other}) = \dfrac{(3 \cdot 2 \cdot 1)(3 \cdot 2 \cdot 1)}{5 \cdot 4 \cdot 3 \cdot 2 \cdot 1} = \dfrac{3}{10}$

(c)

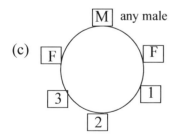

Take male one and seat him at the top. There is a **choice of 3** females next to him 9 (clockwise) and a **choice of 2** females on the other side of that male. This brings us to the adjacent figure which still has 3 vacant seats. Working clockwise, can we place a female in seat 1? No, for then two males would have to sit next to each other in seats 2 and 3. So, there is a **choice of 2** males that we can place in seat 1. But then the remaining female must sit in 2, and the last male must sit in 3. So:

$$Pr(\text{no two males seated next to each other}) = \frac{3 \cdot 2 \cdot 2}{5 \cdot 4 \cdot 3 \cdot 2 \cdot 1} = \frac{1}{10}$$

§1.6 Permutations and Combinations
Page 72

1. $4! = 1 \cdot 2 \cdot 3 \cdot 4 = 24$

3. $P(10, 7) = \dfrac{10!}{(10-7)!} = \dfrac{10!}{3!} = \dfrac{\cancel{3!} \cdot 4 \cdot 5 \cdot 6 \cdot 7 \cdot 8 \cdot 9 \cdot 10}{\cancel{3!}} = 4 \cdot 5 \cdot 6 \cdot 7 \cdot 8 \cdot 9 \cdot 10 = 604{,}800$

5. $C(7, 3) = \dfrac{7!}{(7-3)!3!} = \dfrac{7!}{4!3!} = \dfrac{\cancel{4!} \cdot 5 \cdot 6 \cdot 7}{\cancel{4!}3!} = \dfrac{5 \cdot 6 \cdot 7}{1 \cdot 2 \cdot 3} = 5 \cdot 7 = 35$

7. $P(7, 7) = \dfrac{7!}{(7-7)!} = \dfrac{7!}{0!} = 7! = 1 \cdot 2 \cdot 3 \cdot 4 \cdot 5 \cdot 6 \cdot 7 = 5040$

9. $C(7, 3) - C(7, 4) = \dfrac{7!}{(7-3)!3!} - \dfrac{7!}{(7-4)!4!} = \dfrac{7!}{4!3!} - \dfrac{7!}{3!4!} = 0$

11. $5! = 120$ **13.** $10! = 3{,}628{,}800$ **15.** $P(11, 5) = 55{,}440$ **17.** $P(7, 5) = 2520$

19. $C(27, 4) = 17{,}550$ **21.** $C(52, 13) \approx 6.35 \times 10^{11}$

[**23** and **25**: No unique answers.]

27. (a) Pick 2 from 52, order counts: $P(52, 2) = 2652$.

(b) Grab 2 from 52, order does not count: $C(52, 2) = 1326$.

(c) Two times greater: Every possibility in (b) is counted twice in (a).

29. (Assuming that the order of the toppings is irrelevant)

(a) $C(5, 3) = 10$ (b) $C(5, 1) \overset{\text{or}}{+} C(5, 2) = 5 + 10 = 15$

31.
$\boxed{\begin{array}{l} R_{10} \\ W_8 \\ B_4 \end{array}}$ Grab 3 of the 10 red marbles and 2 of the 8 white marbles:

$C(10, 3) \cdot C(8, 2) = 120 \cdot 28 = 3360$

33.
$\boxed{\begin{array}{l} R_{10} \\ W_8 \\ B_4 \end{array}}$ Grab 1 of the 10 red marbles and 1 of the 8 white marbles and 1 of the 4 blue marbles:

$C(10, 1) \cdot C(8, 1) \cdot C(4.1) = 10 \cdot 8 \cdot 4 = 320$

35.

$\boxed{\begin{array}{l} D_6 \\ \quad N_5 \\ Q_3 \end{array}}$

Grab 3 of the 6 dimes and 2 of the 5 nickels:

$$C(6, 3) \cdot C(5, 2) = 20 \cdot 10 = 200$$

37.

$\boxed{\begin{array}{l} M_7 \\ \quad W_9 \end{array}}$

Grab 3 of the 7 men and 4 of the 9:

$$C(7, 3) \cdot C(9, 4) = 35 \cdot 126 = 4410$$

39.

$\boxed{\begin{array}{l} M_5 \\ \quad B_9 \\ E_{12} \end{array}}$

Grab 2 of the 5 math and 2 of the 9 business courses and 2 of the 12 electives:

$$C(5, 2) \cdot C(9, 2) \cdot C(12, 2) = 10 \cdot 36 \cdot 66 = 23{,}760$$

41.

$\boxed{\begin{array}{l} K_4 \\ \quad Q_4 \end{array}}$

Grab 2 of the 4 Kings and 3 of the 4 Queens:

$$C(4, 2) \cdot C(4, 3) = 6 \cdot 4 = 24$$

43.

$\boxed{\begin{array}{l} C_{13} \\ \quad S_{13} \\ H_{13} \end{array}}$

Grab 2 of 13 Clubs and 1 of the 13 Spades and 2 of the 13 Hearts:

$$C(13, 2) \cdot C(13, 1) \cdot C(13, 2) = 78 \cdot 13 \cdot 78 = 79{,}092$$

45. Pick 4 of 5 flags (can't pick the red) -- order counts: $P(5, 4) = 120$.

47. Pick 4 of 4 flags (can't pick red or yellow) -- order counts: $P(4, 4) = 4! = 24$.

49. Pick 5 of the six flags or 6 of the six flags:

$$P(6, 5) + P(6, 6) = 720 + 720 = 1440$$

51. Possibilities: number of ways of ordering the 6 flags: 6!

Successes: Red on top or red on bottom:

$$1 \cdot 5! + 5! \cdot 1$$

the first flag must be the red and the other 5 can go up in any order.   the last flag must be last and the other 5 can go up in any order $Pr(\text{R on top or bottom}) = \dfrac{1 \cdot 5! + 5! \cdot 1}{6!} = \dfrac{1}{3}$

53. Red white and blue on top in any order

$$3! \cdot 3!$$

and the other three in any order $Pr(\text{RWB on top in any order}) = \dfrac{3!3!}{6!} = \dfrac{1}{20}$

ways of ordering the 6 flags

55. Since order does not "count" in the success, you can do this problem two ways. One way, order counts:

Possibilities when order counts: $P(6,4) = \dfrac{6!}{(6-4)!}$ Successes: $P(5,4)$

The four flags must be chosedn from 5 (no red)

$$Pr(\text{red not hoisted}) = \frac{P(5,4)}{P(6,4)} = \frac{\frac{5!}{(5-4)!}}{\frac{6!}{(6-4)!}} = \frac{\frac{5!}{}}{\frac{6!}{2!}} = \frac{1\cdot2\cdot3\cdot4\cdot5}{3\cdot4\cdot5\cdot6} = \frac{1}{3}$$

Another way, order does not count:

Possibilities when order does not count: $C(6,4)$ Successes: $C(5,4)$

The four flags must be chosedn from 5 (no red)

$$Pr(\text{red not hoisted}) = \frac{C(5,4)}{C(6,4)} = \frac{\frac{5!}{(5-4)!4!}}{\frac{6!}{(6-4)!4!}} = \frac{5!}{(5-4)!4!} \cdot \frac{(6-4)!4!}{6!} = \frac{1\cdot2\cdot3\cdot4\cdot5}{3\cdot4\cdot5\cdot6} = \frac{1}{3}$$

invert and multiply

57. Unlike the situation in **55**, we do not have the "order does not count" option here, as the success is based on an order condition (red flag **on top**).

Possibilities when order counts: $P(6,4)$

Successes: The top flag must be red.

The next three flags are chosen from h remaining 5: $1 \cdot P(5,3)$

$$Pr(\text{red hoisted first}) = \frac{P(5,3)}{P(6,4)} = \frac{60}{360} = \frac{1}{6}$$

59. There is no choice when it comes to the first three flags: Red then White then Blue. The last flag, however, can be any or the3 remaining flags. So:

choices for the fourth flag

$$Pr(\text{R on top then W then B}) = \frac{3}{P(6,4)} = \frac{3}{360} = \frac{1}{120}$$

possibilities

61. Since the success does not specify an order, we can use two approaches. The easier one is when "order does not count." You can then think of the situation as that of an urn containing 2 **S**pecial flags (the red and green), along with 4 **N**ot special flags. Then:

grab 4

S_2 / N_4

Possibilities: $C(6,4)$

Successes: $C(2,2) \cdot C(4,2) = C(4,2)$ $Pr(\text{red and green are hoisted}) = \dfrac{C(4,2)}{C(6,4)} = \dfrac{2}{5}$

grab 2 of the 2 **S** and 2 of the 4 **N**

The harder way (order counts): Possibilities: $P(6, 4)$. It is important to note that the successes constitute a subset of the sample space. So, since order counted in our sample space, it **must also count** when we look for the successes. Let's do so:

We first choose two of the four **positions** to drop the Red and Green flag, with order counting: $P(4, 2)$ [order counts since to drop red on top of green in those **positions** is not the same as placing green on top of red]. Two positions remain and we can place any two of the remaining four flags in those spots, with order counting: $P(4, 2)$.

$$\text{Leading us to: } Pr(\text{red and green are hoisted}) = \frac{P(4, 2) \cdot P(4, 2)}{P(6, 4)} = \frac{2}{5}$$

63. Since order is of no consequence on the success (*red or green are hoisted, possibly both*), we take the easier "order does not count" approach. We are then looking at an **urn problem** which consists of 2 **S**pecial flags (red and green) and 4 **N**ot special flags:

grab 4

S_2
N_4

Possibilities: $C(6, 4)$

Successes: $C(2, 2) \cdot C(4, 2) + C(2, 1) \cdot C(4, 3)$

grab 2 of the 2 **S** and 2 of the 4 **N**   OR   1 of the 2S and 3 of the four N

$$\text{Leading us to: } Pr(\text{red or green, possily both}) = \frac{C(2, 2)C(4, 2) + C(2, 1)C(4, 3)}{C(6, 4)} = \frac{14}{15}$$

65.
R_9
W_7
B_5

4

grab 4 of the non-red

$$Pr(\text{all R}) = \frac{C(12, 4)}{C(21, 4)} = \frac{11}{133} \approx 0.083$$

grab 4 from 21

67.
R_9
W_7
B_5

4

grab 4 of the 9 red or 4 of the 7 white or 4 of the 5 blue

$$Pr(\text{all of the same color}) = \frac{C(9, 4) + C(7, 4) + C(5, 4)}{C(21, 4)} = \frac{166}{5985} \approx 0.028$$

grab 4 from 21

69. If each of the three colors are drawn than you must grab a red (1 of 9), a white (1 of 7), and a blue (1 of 5) — along with another color (1 of the other colors). This, however would be **WRONG**:

R_9
W_7
B_5

4

Grab a red and three of the other or a white and three of the other or a blue and 3 of the other

$$Pr(\text{one of each}) = \frac{C(9, 1)C(12, 3) + C(7, 1)C(14, 3) + C(5, 1)C(16, 3)}{C(21, 4)}$$

WRONG

You will see that something must be wrong if you calculated the above expression, for it turns out to have a value greater than 1. The problem is that in the above, events are counted more than once. Picking a red and a white and two blues, for example, is counted both in $C(9, 1)C(12, 3)$ and in $C(7, 1)C(14, 3)$ (think about it). Here is the correct answer:

grab 2 reds a white and a blue **or** two whites a red and a blue **or** two blues a red and a white

$$\frac{C(9, 2)C(7, 1)C(5, 1) + C(7, 2)C(9, 1)C(5, 1) + C(5, 2)C(9, 1)C(7, 1)}{C(21, 4)} = \frac{9}{19} \approx 0.47$$

Note that there is no multiple counting in the above expression.

71. $\boxed{W_9 \;\; M_6}$ $\nearrow 5$

grab 3 of the women and 2 of the men

$$Pr(\text{exactly 3 W}) = \frac{C(9, 3)C(6, 2)}{C(15, 5)} = \frac{60}{143} \approx 0.42$$

grab 5 from 15

73. $\boxed{W_9 \;\; M_6}$ $\nearrow 5$

not: 4 women and 1 man or 5 women and no men

$$Pr(\text{at most 3 W}) = 1 - \frac{C(9, 4)C(6, 1) + C(9, 5)C(6, 0)}{C(15, 5)} = \frac{101}{143} \approx 0.71$$

75. While an order is imposed on the five prizes, order does not play a role in defining a success: *exactly three of the girls receive an award* (no order). That being the case the solution of Exercise 71 is a solution for this problem. If you wish, however, you can focus on an "ordered' sample space to arrive at the same answer, but it is more complicated:

$\nearrow 5$ (order counts)

$\boxed{G_9 \;\; B_6}$

A success: **first** second **third fourth** fifth
　　　　　G   B   G   G   B

grab the three prizes that are going to the girls
pick the three girls that get those prizes
(order counts)

$$Pr(\text{exactly 3 girls get a prize}) = \frac{C(5, 3)P(9, 3)P(6, 2)}{P(15, 5)} = \frac{60}{143} \approx 0.42$$

pick any 5 of the 15 children (order counts)

pick the two boys for the remaining two prizes

77. Pick three of the girls to win the **specified** first, second, and fifth prizes. **Choices:** $P(9, 3)$.
Two of the 6 boys **must** win the third and fourth prizes. **Choices:** $P(6, 2)$. Conclusion:

$$Pr(\text{girls win the first, second, and third prizes; boys the rest}) = \frac{P(9, 3)P(6, 2)}{P(15, 5)} = \frac{6}{143} \approx 0.042$$

79. This is an urn problems consisting of 4 King marbles and 48 Non-King marbles:

$\boxed{K_4 \;\; N_{48}}$ $\nearrow 2$

$$P(\text{no K}) = \frac{C(4, 0)C(48, 2)}{C(52, 2)} = \frac{188}{221} \approx 0.85$$

81. An urn problem consisting of 13 Club-marbles and 39 Non-Club-marbles:

$$P(\text{all C}) = \frac{C(13, 5)C(39, 0)}{C(52, 5)} = \frac{33}{66{,}640} \approx 0.0005$$

with box: C_{13} N_{39} (arrow 5, down 1)

83. Grab 3 of the 4 Kings and 2 of the 4 Queens (and none of the Non-King-Queen cards:

$$P(3K \text{ and } 2Q) = \frac{C(4, 3)C(4, 2)}{C(52, 5)} = \frac{1}{108{,}290} \approx 0.000009$$

with box: K_4 Q_4 N_{44} (arrow 5)

85. A "reasonable" **wrong** answer: Choose 1 of the 13 types (say Kings) and choose three of that type; then go ahead and choose one of the remaining 12 types (say Queens) and choose 2 of that type; finally chose a fifth card that is neither of the two chosen types (choice of 1 out of 44). Leading us to the following **wrong** answer:

$$Pr(\text{two pairs}) \underset{\text{WRONG}}{=} \frac{C(13, 1)C(4, 2)C(12, 1)C(4, 2)C(44, 1)}{C(52, 5)} \approx 0.096$$

The problem with the above argument is that each two pair hand is counted twice. Here is a correct solution:

grab two of the 13 types   grab 2 of one of those types an 2 of the other   grab 1 of neither type

$$Pr(\text{two pairs}) = \frac{C(13, 2)C(4, 2)C(4, 2)C(44, 1)}{C(52, 5)} = \frac{198}{4165} \approx 0.048 \text{ (half of the above wrong answer)}$$

87. There are 4 royal flushes: 10-J-Q-K-A of Hearts; of Clubs; of Diamonds; and of Spades. So:

$$Pr(\text{royal flush}) = \frac{4}{C(52, 5)} = \frac{1}{649{,}740} \approx 0.0000015$$

the highest poker hand -- cannot be beat.

89. If asked to write down a straight you may go with: 2-3-4-5-6; or perhaps: 9-10-J-Q-K. The first choice is to pick the starting card ("2" in the one case, and "9" in the other). So, you have a choice of 10 for starting the straight (you cannot start it with a Jack -- why not?). But if you go with the 2-3-4-5-6 choice, you still have other choices: choose one of the four 2's; one of the four 3's; one of the four 4's; one of the four 5's; and one of the four 6's. Putting this together we conclude that there are $10 \cdot 4^5$ different straights. But our straights are not be of one suit and there are $10 \cdot 4$ of them (10 choices of where they start and 4 choices for the suit). So:

$$Pr(\text{straigh that is not a straight-flush or a royal flush}) = \frac{10 \cdot 4^5 - 10 \cdot 4}{C(52, 5)} = \frac{5}{1274} \approx 0.0039$$

91. There are only four successes: all of the 13 cards are Hearts; or Clubs; or Spades; or diamonds. So:

$$Pr(\text{same suit}) = \frac{4}{C(52, 13)} = \frac{1}{15{,}875{,}3389{,}900} \approx 6.3 \times 10^{-12} \text{ (not bloody likely)}$$

93. Let's think in terms of urns: an urn containing 16 Face-or-Ace marbles (FA marbles), and 32 Other marbles. Then:

$FA_{16} \; O_{32}$ → 13

$$P(\text{all } FA) = \frac{C(16, 13)C(32, 0)}{C(52, 13)} = \frac{1}{1,133,952,785} \approx 8.8 \times 10^{-10}$$

95. (a) Since the box will not be shipped only if more than one of the five tested calculators is defective, and since the box of 24 calculators contains but one defective calculator, the probability is 1 that the box will be shipped.

(b) Lets consider the urn consisting of 5 Defective marbles and 19 Good marbles. The box will be shipped if of the 5 marbles grabbed at least 4 are Good; which is to say that only 0 or 1 Defective marble is chosen:

$D_5 \; G_{19}$ → 5

$$Pr(\text{shipped}) = \frac{C(5, 0)C(19, 5) + C(5, 1)C(19, 4)}{C(24, 5)} = \frac{1292}{1771} \approx 0.73$$

(c) $D_{19} \; G_5$ → 5

$$Pr(\text{shipped}) = \frac{C(19, 0)C(5, 5) + C(19, 1)C(5, 4)}{C(24, 5)} = \frac{4}{1771} \approx 0.0023$$

97. (a) Pick any 3 or the 7 horses to come in win-place-or show (order counts): $P(7, 3) = 210$.

(b) Since there are 7 horses, the probability that yours will come in first is $\frac{1}{7}$.

(c) There is a probability of $\frac{1}{7}$ that your horse will come in first, a $\frac{1}{7}$ probability that it will come in second, and a $\frac{1}{7}$ probability that it will come in third. So:

$$Pr(\text{first or second or third}) = 3 \cdot \frac{1}{7}.$$

99. (a) Since you bet on 3 horses and since 7 are in the race, the probability that one of your horses will come in first is $\frac{3}{7}$.

(b) You are not concerned on which of your horses comes in, other than all three have to come in first, second, or third. Let's think in terms of an urn consisting of 3 Special marbles (your three horses) and 4 four Other marbles. Grabbing three of those marbles (the marble coming in first, second, and third) we have:

$S_3 \; O_4$ → 3

$$Pr(\text{all 3 are } S) = \frac{C(3, 3)C(4, 0)}{C(7, 3)} = \frac{1 \cdot 1}{C(7, 3)} = \frac{1}{35}$$

(c) $S_3 \; O_4$ → 3

$$Pr(2S \text{ and } 1O) = \frac{C(3, 2)C(4, 1)}{C(7, 3)} = \frac{12}{35}$$

(d) $S_3 \; O_4$ → 3

$$Pr(1S \text{ and } 2O) = \frac{C(3, 1)C(4, 2)}{C(7, 3)} = \frac{18}{35}$$

(e) $\boxed{\begin{array}{c} S_3 \\ O_4 \end{array}} \nearrow 3$ $Pr(\text{at least one } S) = 1 \overset{\textbf{not no-S}}{\underset{\downarrow}{-}} \dfrac{C(3,0)C(4,3)}{C(7,3)} = \dfrac{31}{35}$

(f) $\boxed{\begin{array}{c} S_3 \\ O_4 \end{array}} \nearrow 3$ $Pr(\text{at least two } S) = \dfrac{C(3,2)C(4,1)+C(3,3)C(4,0)}{C(7,3)} = \dfrac{13}{35}$

101. (a) There are as many possible winning numbers as there are of grabbing 5 numbers from 50. There is only one way to match all of the 5 numbers. So:

$$Pr(\text{winning a million}) = \frac{1}{C(50,5)} = \frac{1}{2{,}118{,}760} \approx 0.00000047$$

(b) Suppose, for definiteness, that the winning numbers are 3 7 11 21 42. To win the $10,000 you must miss exactly one of those five numbers; for example: 3 **43** 11 21 42. We had a **choice of 5** numbers to change (we changed the 7). We changed it to a 43, but could have replaced it with any of the 50 numbers excluding the numbers 3, 11, 21, 7, and 42 — **a choice of 45**. Consequently:

$$Pr(\text{winning } \$10{,}000) = \frac{5 \cdot 45}{C(50,5)} = \frac{45}{423{,}752} \approx 0.00011$$

(c) To win nothing is **not** to win something. So:

$$Pr(\text{win nothing}) = 1 - [\text{answer to (a)} + \text{answer to (b)}] = 1 - \left[\frac{1}{C(50,5)} + \frac{5 \cdot 45}{C(50,5)}\right] \approx 0.99989$$

103. (a) The are as many possible ways of matching the 10 questions to the 10 answers as there are ways of ordering the 10 questions (if Question 1 turns out to be third, then link it with Answer 3; and so on). So there are 10! ways of linking the 10 questions with the 10 answers. Since only one of these ways will result in receiving 100 points (every question must be linked with its answer):

$$Pr(100 \text{ points}) = \frac{1}{10!} = \frac{1}{3{,}628{,}800} \approx 0.00000027$$

(b) To get 90 pints you would need to get exactly one question wrong. This cannot happen; for if you link a question with a wrong answer, then the question that should be linked to that answer can no longer be accurately linked. So, $Pr(90 \text{ points}) = 0$.

(c) To get 80 points, exactly two of the 10 question must be assigned a wrong answers. Choices for grabbing two of the 10 questions: $C(10,2)$. Having chosen the two that will be linked to incorrect answers, are there any other choices? No, for each of the two **must** be linked with the correct answer of the other, and each of the remaining 8 questions **must** be linked with its correct answer. So:

$$Pr(80 \text{ points}) = \frac{C(10,2)}{10!} = \frac{1}{80640} \approx 0.000012$$

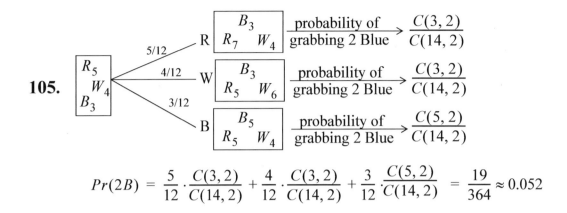

$$Pr(2B) = \frac{5}{12} \cdot \frac{C(3,2)}{C(14,2)} + \frac{4}{12} \cdot \frac{C(3,2)}{C(14,2)} + \frac{3}{12} \cdot \frac{C(5,2)}{C(14,2)} = \frac{19}{364} \approx 0.052$$

107. (a) Let's place a bar on one of the two O's in the word OHIO: $\bar{O}$HIO . We now have four distinguishable letters and can order them in 4! ways, with the string H$\bar{O}$OI for example, being different from the string HO$\bar{O}$I . But if you pull the bar away, those two strings, are no longer different, one from the other. It follows that the 4! strings that are generated when the bar is imposed to distinguish one O from the other, must be divided by 2 once the bar is removed. So, the number of different four letter character strings that are generated by the letters in the word OHIO is $\frac{4!}{2} = 12$.

(b) In CALIFORNIA there are two A's, and two I's. If we reason as in (a) and place a bar over one of the A's and a bar over one of the I's, we could then generate 10! different strings using the ten distinguishable letters in C$\bar{A}$L$\bar{I}$FORNIA . But once we pull away the bars we see that each of the 9! string that are generated when the bars are imposed is actually counted 4 times (twice because of the two A's and twice because of the two I's). So, the number of different 10 letter character strings that are generated by the letters in the word CALIFORNIA is $\frac{10!}{4} = 907{,}200$.

(c) In MISSISSIPPI there are two I's, four S's, and two P's. If you paint the I's with two different colors, the S's with four different colors, and the P's with two different colors, then you are looking at a collection of 11 distinguishable letter, which can generate 11! different strings. But once we remove the imposed paint, we see that each of the 11! strings were counted $2 \cdot 4! \cdot 2$ times [twice (or 2!) for the two I's, 4! for the number of ways the four S's can be arranged (when distinguished one from the other), and 2 (or 2!) for the two P's. So, the number of different 11 letter character strings that are generated by the letters in the word MISSISSIPPI is $\frac{11!}{2!4!2!} = 34{,}650$.

109. The number of ways of grabbing 52 objects from 100 equals the number of ways of leaving 48 of the 100 in place. In general: $C(n, r) = C(n, n - r)$.

§1.7 Bernoulli Trials
Page 84

1. (a) This is not a Bernoulli Trials situation, since the 2's must occur on the first 3 rolls.)
Applying Theorem 1.6, page 33, we have:

$$Pr\left[2\ 2\ 2\ \cancel{2}\cancel{2}\cancel{2}\right] = \frac{1}{6}\cdot\frac{1}{6}\cdot\frac{1}{6}\cdot\frac{5}{6}\cdot\frac{5}{6}\cdot\frac{5}{6} = \left(\frac{1}{6}\right)^3\left(\frac{5}{6}\right)^3 \approx 0.0027$$

(b)

n	p	r
6	$\frac{1}{6}$	3

$$Pr(2 \text{ is rolled exactly 3 times}) = C(6,3)\left(\frac{1}{6}\right)^3\left(\frac{5}{6}\right)^3 \approx 0.054$$

(c)

n	p	r
6	$\frac{1}{6}$	5 or 6

$$Pr(2 \text{ is rolled at least 5 times})$$
$$= C(6,5)\left(\frac{1}{6}\right)^5\left(\frac{5}{6}\right)^1 + C(6,6)\left(\frac{1}{6}\right)^6\left(\frac{5}{6}\right)^0 \approx 0.00066$$

(d)

n	p	r
6	$\frac{1}{6}$	not 6

$$Pr(2 \text{ is rolled at most 5 times}) = 1 - C(6,6)\left(\frac{1}{6}\right)^6\left(\frac{5}{6}\right)^0 \approx 0.9998$$

(e)

n	p	r
6	$\frac{2}{6}$	4

$$Pr(1 \text{ or 2 is rolled 4 times}) = C(6,4)\left(\frac{2}{6}\right)^4\left(\frac{4}{6}\right)^2 \approx 0.082$$

(f) (This is not a Bernoulli Trials situation, since the success is specified to occur on two specified rolls of the die: first and fifth roll). Applying Theorem 1.6, page 33, we have:

$$Pr\left(\left[<3\ \cancel{\cancel{3}}\ \cancel{\cancel{3}}\ \cancel{\cancel{3}}\ <3\ \cancel{\cancel{3}}\right]\right) = \frac{2}{6}\cdot\frac{4}{6}\cdot\frac{4}{6}\cdot\frac{4}{6}\cdot\frac{2}{6}\cdot\frac{4}{6} = \left(\frac{2}{6}\right)^2\left(\frac{4}{6}\right)^4 \approx 0.022$$

(g) Again we are confronted with a non-Bernoulli trial, as the two success do not have complete freedom as to where they can occur (one must follow the other). Going directly to the probability definition, we first determine the number of possible outcomes when rolling a die 6 times, namely 6^6 possibilities. Now, here is a particular success: 5 1 1 3 3 6. Our main concern was to get two 1's next to each other, and decided to put the first 1 in the second position: **5 choices** (could not have placed the first 1 at the end). There is but one choice as to what sits to the right of that initial 1, for it must be a 1. However, in the remaining 4 positions we could put any number between 2 and 6 (can't have a third 1); 5 choices 4 times, for a total of **5^4 choices**. Conclusion:

$$Pr(1 \text{ is rolle twice, with one righ after the other }) = \frac{5\cdot 5^4}{6^6} \approx 0.067$$

(h) To be a Bernoulli trial the success has to be associated with each trial. This is not the situation here, for the specified success, *the same number is rolled each time,* in not linked with each roll of the die but rather with the roll of the six dice. Noting that *the same number is rolled each time* can happen in six ways (the number 1 is rolled each time (a Bernoulli situation), or the number 2 is rolled each time, or..., or the number 6 is rolled each time), and that each of these six ways is as likely to occur as any other, we can get to our answer by simply multiplying the probability that a 1 being rolled each time by 6. Let's do it:

n	p	r
6	$\frac{1}{6}$	6

$$Pr(1 \text{ is rolled 6 times}) = C(6, 6)\left(\frac{1}{6}\right)^6\left(\frac{5}{6}\right)^0 = \left(\frac{1}{6}\right)^6$$

So: $Pr(\text{same number each time}) = 6 \cdot \left(\frac{1}{6}\right)^6 = \left(\frac{1}{6}\right)^5 \approx 0.00012$

Another way: Roll a dice 6 times. Number of possible outcomes 6^6. Define a success to be: same number is rolled each time. Number of successes: 6 (1 or 2 or 3 or 4 or 5 or 6 could be rolled each time). And again we have:

$$Pr(\text{same number each time}) = \frac{6}{6^6} = \left(\frac{1}{6}\right)^5$$

3. (a) (Bernoulli) Since you replace the slip each time, you are indeed conducting the same experiment 9 times, with probability of success $\frac{1}{9}$ (there are 9 numbers in the hat). Thus:

n	p	r
9	$\frac{1}{9}$	2

$$Pr(5 \text{ exactly two times}) = C(9, 2)\left(\frac{1}{9}\right)^2\left(\frac{8}{9}\right)^7 \approx 0.19$$

(b) (Not Bernoulli: The same experiment is not conduction each time — the hat gets "lighter" each time.) Since the slip is not replaced, there is 0 probability that the number 5 is drawn twice.

5. (a)

n	p	r
7	0.67	7

$$Pr(\text{all accept}) = C(7, 7)(0.67)^7(0.33)^0 = (0.67)^7 \approx 0.61$$

(b)

n	p	r
7	0.67	0

$$Pr(\text{none accept}) = C(7, 7)(0.67)^0(0.33)^7 = (0.33)^7 \approx 0.00043$$

(c)

n	p	r
7	0.67	1

$$Pr(1 \text{ accepts}) = C(7, 1)(0.67)^1(0.33)^6 = (0.33)^7 \approx 0.0061$$

(d)

n	p	r
7	0.67	not 0

$$Pr(\text{at lest 1}) = 1 - Pr(\text{none}) = 1 - C(7, 7)(0.67)^0(0.33)^7 \approx 0.99957$$

7. (a)

n	p	r
4	0.95	4

$Pr(\text{all four}) = C(4,4)(0.95)^4(0.33)^0 = (0.95)^4 \approx 0.81$

(b)

n	p	r
4	0.95	0

$Pr(\text{none}) = C(7,0)(0.95)^0(0.05)^4 = (0.05)^4 \approx 0.0000063$

(c)

n	p	r
4	0.95	2

$Pr(\text{exactly } 2) = C(7,2)(0.95)^2(0.05)^2 \approx 0.014$

(d)

n	p	r
7	0.95	not 0 or 1

$Pr(\text{at least } 2)$

$= 1 - [C(7,0)(0.95)^0(0.05)^4 + C(7,1)(0.95)^1(0.05)^3] \approx 0.9992$

9. (a)

n	p	r
5	0.57	4

$Pr(\text{exactly } 4) = C(5,4)(0.57)^4(0.43)^1 \approx 0.23$

(b)

n	p	r
5	0.57	4 or 5

$Pr(4 \text{ or } 5) = C(5,4)(0.57)^4(0.43)^1 + C(5,5)(0.57)^5(0.43)^0 \approx 0.29$

(c)

n	p	r
5	0.57	not 5

$Pr(\text{at most } 4) = 1 - C(5,5)(0.57)^5(0.43)^0 \approx 0.94$

11. The first order of business is to find the probability, p, that a bridge hand (13 cards) contains all four aces — an urn problem, with the urn consisting of 4 **Aces** and 48 **Non-Aces**:

grab 4 of the A's and 9 of the N's

$$p = Pr(4 \text{ Aces}) = \frac{C(4,4) \cdot C(48,9)}{C(52,13)}$$

(*)

grab 13 of 52 cards

Then:

(a)

n	p	r
100	(*)	1

$Pr(\text{exactly } 1) = C(100,1)p^1[1-p]^{99}$

see (*): $= 100 \cdot \left[\dfrac{C(4,4) \cdot C(48,9)}{C(52,13)}\right]\left[1 - \dfrac{C(4,4) \cdot C(48,9)}{C(52,13)}\right]^{99} \approx 0.20$

(b)

n	p	r
100	(*)	0

$Pr(\text{none}) = C(100,0)p^0[1-p]^{100} = \left[1 - \dfrac{C(4,4) \cdot C(48,9)}{C(52,13)}\right]^{100} \approx 0.77$

(*)

(c)

n	p	r
100	(*)	not 0

$Pr(\text{at least } 1) = 1 - Pr(\text{none}) = 1 - \left[1 - \dfrac{C(4,4) \cdot C(48,9)}{C(52,13)}\right]^{100} \approx 0.23$

from (b)

13. (a)

n	p	r
12	$\frac{1}{9}$	3

$Pr(\text{win 3 pennies}) = C(12, 3)\left(\frac{1}{9}\right)^3\left(\frac{8}{9}\right)^9 \approx 0.10$

(b)

n	p	r
12	$\frac{1}{9}$	0,1, or 2

$Pr(\text{win less than 3 pennies})$

$= C(12, 0)\left(\frac{1}{9}\right)^0\left(\frac{8}{9}\right)^{12} + C(12, 1)\left(\frac{1}{9}\right)^1\left(\frac{8}{9}\right)^{11} + C(12, 2)\left(\frac{1}{9}\right)^2\left(\frac{8}{9}\right)^{10} \approx 0.86$

15. The main challenge is to determine the probability of drawing a 5 at the end of the journey. In the diagram below F denotes the Five-marbles in the hat (urn), and N the Non-Five-marbles. In particular, at the far left we see that the hat A originally has one five and 8 non-fives. We are interested in the probability p of arriving at an F-leaf (for drawing the number Five) at the far right. The tree diagram is not as intimidating as it might at first appear (one can only hope). Consider, for example, the development leading to the far-right top F-leaf, labeled numbered (1). To get to that leaf, an F leaf had to be drawn initially from Hat A and placed in Hat B. At this point, as is indicated in the diagram, A has no F-marbles and 8 N-marbles; while B, having gained an F-marble now has 2 F and 8 N marbles. Continuing the good fight, in order to get to top-right-F leaf, an F marble was drawn from B and dumped into A; resulting in A returning to its initial state; after which we can move to the leaf F in question.

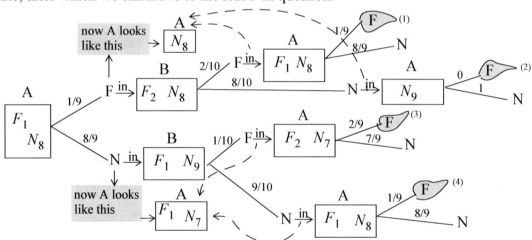

So many words in an attempt to describe what's going on in the figure; which really speaks for itself. And if we listen carefully, it is telling us that the probability p of ending up with an F (the number 5) at the end of the journey is equal to that of ending up with an F at the beginning of the journey; namely $p = 1/9$; for:

$$Pr(\text{drawing a five at end}) = \overset{(1)}{\frac{1}{9}\cdot\frac{2}{10}\cdot\frac{1}{9}} + \overset{(2)}{\frac{1}{9}\cdot\frac{8}{10}\cdot 0} + \overset{(3)}{\frac{8}{9}\cdot\frac{1}{10}\cdot\frac{2}{9}} + \overset{(4)}{\frac{8}{9}\cdot\frac{9}{10}\cdot\frac{1}{9}} = \frac{1}{9}$$

Since the probability of winning a penny in this exercise is exactly the same as it was in Exercise 13, the answers to part (a) and (b) here are identical to those appearing in Exercise 13 above:

(a)

n	p	r
12	$\frac{1}{9}$	3

$Pr(\text{win 3 pennies}) = C(12, 3)\left(\frac{1}{9}\right)^3\left(\frac{8}{9}\right)^9 \approx 0.10$

(b)

n	p	r
12	$\frac{1}{9}$	0,1, or 2

$Pr(\text{win less than 3 pennies})$

$= C(12, 0)\left(\frac{1}{9}\right)^0\left(\frac{8}{9}\right)^{12} + C(12, 1)\left(\frac{1}{9}\right)^1\left(\frac{8}{9}\right)^{11} + C(12, 2)\left(\frac{1}{9}\right)^2\left(\frac{8}{9}\right)^{10} \approx 0.86$

17. (a) We are told that $p = 0.01\% = 0.0001$ is the probability that a moving violation has been administered on **any give day of the year**. The key to the solution is to realize that, in the Bernoulli formula, n is the number of days in a year: $n = 365$. We can then turn to the formula to determine the probability that the company will be willing to renew an individuals policy (not 2 or more violations):

n	p	r
365	0.01%	0 or 1

$Pr(\text{willing to renew a policy})$:

$$C(365, 0)(0.0001)^0(1 - 0.0001)^{365} + C(365, 1)(0.0001)^1(1 - 0.0001)^{364}$$
$$\approx 0.9994$$

(b) We are no longer looking at one individual but, rather, at 93 individuals; and are interested in determining the probability that either 92 or all 93 of their policies will be renewed. So, in the Bernoulli formula: $n = 93$ and $r = 93$ or 92. But what is p? It is the probability that the company will be willing to renew the policy of an individual — the answer to (a) (which is approximately 0.9994):

n	p	r
93	0.9994	93 or 92

$Pr(\text{willing to renew at least 92 of 93 policy holders})$:

$$C(93, 93)(0.9994)^{93}(1 - 0.9994)^0 + C(93, 92)(0.9994)^{92}(1 - 0.9994)^1$$

see (a)

$$\approx 0.9983$$

§1.8 Expected Value and Decision Making
Page 97

1.

Children	0	1	2	3	4	5	6	7
Frequency	25	53	71	18	5	2	0	1

sum $\to 25 + 53 + 71 + 18 + 5 + 2 + 0 + 1 = 175$

posible # of children

$$E(\#\text{ of chlidren}) = 0 \cdot \frac{25}{175} + 1 \cdot \frac{53}{175} + 2 \cdot \frac{71}{175} + 3 \cdot \frac{18}{175} + 4 \cdot \frac{5}{175} + 5 \cdot \frac{2}{175} + 6 \cdot \frac{0}{175} + 7 \cdot \frac{1}{175} \approx 1.63$$

probability of occurance

3. From the sample space:

HHH
HHT HTH THH
HTT THT TTH
TTT

we have:

number of T	Probability
0	1/8
1	3/8
2	3/8
3	1/8

So, as might be expected: $E(\#\text{ of T}) = 0 \cdot \frac{1}{8} + 1 \cdot \frac{3}{8} + 2 \cdot \frac{3}{8} + 3 \cdot \frac{1}{8} = \frac{12}{8} = \frac{3}{2}$.

5. The possible value of women are 0, 1, 2, and 3. As for their probability of occurrence:

$$\nearrow^{3}$$

$$\boxed{\begin{array}{c} M_5 \\ W_7 \end{array}}$$

$$Pr(\text{no } W) = \frac{C(7,0)C(5,3)}{C(12,3)} \qquad Pr(\text{one } W) = \frac{C(7,1)C(5,2)}{C(12,3)}$$

So:

$$Pr(\text{two } W) = \frac{C(7,2)C(5,1)}{C(12,3)} \qquad Pr(\text{three } W) = \frac{C(7,3)C(5,0)}{C(12,3)}$$

$$E(\#\text{ of Women}) = 0 \cdot \frac{C(7,0)C(5,3)}{C(12,3)} + 1 \cdot \frac{C(7,1)C(5,2)}{C(12,3)} + 2 \cdot \frac{C(7,2)C(5,1)}{C(12,3)} + 3 \cdot \frac{C(7,3)C(5,0)}{C(12,3)} \approx 1.75$$

An interesting observation: The above answer equals 3 times the probability of drawing a W from the urn: $3\left(\frac{7}{12}\right)$.

7. The possible values of clubs are 0, 1, 2, and 3. As for their probability of occurrence:

$$\nearrow^{3}$$

$$\boxed{\begin{array}{c} C_{13} \\ N_{39} \end{array}}$$

$$Pr(\text{no } C) = \frac{C(13,0)C(39,3)}{C(52,3)} \qquad Pr(\text{one } C) = \frac{C(13,1)C(39,2)}{C(52,3)}$$

So:

$$Pr(\text{two } C) = \frac{C(13,2)C(39,1)}{C(52,3)} \qquad Pr(\text{three } C) = \frac{C(13,3)C(39,0)}{C(52,3)}$$

$$E(\#\text{ of Clubs}) = 0 \cdot \frac{C(13,0)C(39,3)}{C(52,3)} + 1 \cdot \frac{C(13,1)C(39,2)}{C(52,3)} + 2 \cdot \frac{C(13,2)C(39,1)}{C(52,3)} + 3 \cdot \frac{C(13,3)C(39,0)}{C(52,3)} \approx 0.75$$

An interesting observation: The above answer equals 3 times the probability of drawing a C from the above urn: $3\left(\frac{13}{52}\right)$.

9. A glance at the sample space of Figure 1.2, page 3, reveals the fact that $Pr(\text{doubles}) = \frac{6}{36}$. It follows that the probability of not rolling doubles is $1 - \frac{6}{36} = \frac{30}{36}$. So:

$$E(\text{Winnings}) = 10 \cdot \frac{6}{36} - 1.50 \cdot \frac{30}{36} = \$0.42$$

11. Let's figure out the probabilities associated with winning $5 and $2 and losing $3.20:

$$\nearrow^{1}$$

$$\boxed{\begin{array}{c} R_5 \\ B_3 \, W_2 \end{array}}$$

$$Pr(\text{win } \$5) = Pr(W) = \frac{2}{10} \qquad Pr(\text{win } \$2) = Pr(B) = \frac{3}{10}$$

$$Pr(\text{lose } \$3.20) = Pr(R) = \frac{5}{10}$$

So: $E(\text{Winnings}) = 5 \cdot \frac{2}{10} + 2 \cdot \frac{3}{10} - 3.2 \cdot \frac{5}{10} = 0$.

13. To win $10 you have to draw a pair, and to win $1 you need to end up with two cards of the same suit. Figuring out the corresponding probabilities we have:

choice of 13 types (Aces, or Kings, or, ...)   grab 2 of that type

$$Pr(\text{pair}) = \frac{13\,C(4,2)}{C(52,2)}$$

grab 2 of the 52 cards

choice of 4 suits   grab 2 of that suit

$$Pr(\text{same suit}) = \frac{4\,C(13,2)}{C(52,2)}$$

You will lose a buck if you do not draw a pair or draw two cards of the same suit, and the probability of that happening is 1 minus the probability of drawing a pair or two of the same suit; namely: $1 - \left(\dfrac{13\,C(4,2)}{C(52,2)} + \dfrac{4\,C(13,2)}{C(52,2)}\right)$. So:

$$E(\text{Winnings}) = 10 \cdot \frac{13\,C(4,2)}{C(52,2)} + 1 \cdot \frac{4\,C(13,2)}{C(52,2)} - 1 \cdot \left[1 - \left(\frac{13\,C(4,2)}{C(52,2)} + \frac{4\,C(13,2)}{C(52,2)}\right)\right] \approx \$0.12$$

15. This is just like Example 1.47, but with different numbers. Here we have:

$$E(\text{Winnings}) = 498 \cdot \frac{1}{1000} + 248 \cdot \frac{2}{1000} + 98 \cdot \frac{3}{1000} - 2 \cdot \frac{994}{1000} = -\$0.70$$

17. To win a million, all five of your numbers, chosen from 50, must match the drawn five numbers, and here is the probability of that happening: $\dfrac{1 \leftarrow \text{but 1 success}}{C(50,5) \leftarrow \text{grab 5 from 50}}$

To win $10,000 exactly four of your five numbers must match the drawn numbers. The probability of this happening turns out to be $\dfrac{5 \cdot 45}{C(50,5)}$ [see solution of Exercise 101 (b) appearing on page 29]. Bringing us to:

$$E(\text{Winnings}) = 1,000,000 \cdot \frac{1}{C(50,5)} + 10,000 \cdot \frac{5 \cdot 45}{C(50,5)} - 2 = -\$0.47$$

19. Turning directly to Theorem 1.13 we have: $E(\text{number of quitters}) =$

probability of quitting   number of smoker

$0.37 \cdot 100 = 37$.

21. From the sample space

HH
HT TH
TT

, or from Theorem 1.6, page 33, we conclude that there is a $\dfrac{1}{4}$ probability of flipping two Tails when a coin is flipped twice. Applying Theorem 1.13 we have:

probability of flipping TT   number of flips

$$E(\text{number of HH}) = \frac{1}{4} \cdot 40 = 10$$

23. Turning directly to Theorem 1.13 we have: $E(\text{number of A's}) = \underset{\underset{\text{probability of A}}{\downarrow}}{0.14} \cdot \underset{\underset{\text{number of students}}{\downarrow}}{25} = 3.5$.

25. (a) Turning to:

Face rolled	1	2	3	4	5	6
Probability	1/12	2/12	3/12	3/12	1/12	2/12

we have: $E(\text{roll of die}) = 1 \cdot \dfrac{1}{12} + 2 \cdot \dfrac{2}{12} + 3 \cdot \dfrac{3}{12} + 4 \cdot \dfrac{3}{12} + 5 \cdot \dfrac{1}{12} + 6 \cdot \dfrac{2}{12} = \dfrac{43}{12} \approx 3.58$.

(b) Applying Theorem 1.13: $E(\text{number of 3's}) = \underset{\underset{\text{probability of 3}}{\downarrow}}{\dfrac{3}{12}} \cdot \underset{\underset{\text{number of students}}{\downarrow}}{100} = 25$.

27. (a) The probability that you hit the bulls-eye, given that the dart will randomly hit somewhere in the board, is the ratio of the area of the bull's-eye to that of the board; namely: $\dfrac{\pi 1^2}{\pi 3^2} = \dfrac{1}{9}$. To get 50 points, the dart has to land within the 2 inch circle but outside of the bull's-eye: $\dfrac{\pi 2^2 - \pi 1^2}{\pi 3^2} = \dfrac{1}{3}$.

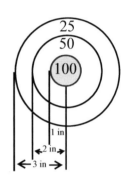

Similarly, the probability of getting 25 points is $\dfrac{\pi 3^2 - \pi 2^2}{\pi 3^2} = \dfrac{5}{9}$.

So: $E(\text{number of points}) = 100 \cdot \dfrac{1}{9} + 50 \cdot \dfrac{1}{3} + 25 \cdot \dfrac{5}{9} = \dfrac{125}{3}$

(b) Applying Theorem 1.13: $E(\text{number of bull's-eye}) = \underset{\underset{\text{probability of bull's-eye}}{\downarrow}}{\dfrac{1}{9}} \cdot \underset{\underset{\text{number of tosses}}{\downarrow}}{30} = \dfrac{10}{3}$.

29. For each $1 \le i \le 100$, let X_i be the random variable that assigns the value 1 if person i draws his or her key from the hat, and the value 0 if not. Here is the expected value of each X_i:

$E(X_i) = 0 \cdot \dfrac{99}{100} + 1 \cdot \dfrac{1}{100} = \dfrac{1}{100}$. Let S denote the random variable which assigns the total number of people drawing their own key: $S = \displaystyle\sum_{i=1}^{100} X_i$. Applying the sum theorem, we find that:

$E(S) = \displaystyle\sum_{i=1}^{100} E(X_i) = \sum_{i=1}^{100} \dfrac{1}{100} = 1$. Conclusion: On the average, exactly one person will draw his or her key from the hat (independently of the number of people involved!).

31. For each $1 \leq i \leq n$, let X_i be the random variable that assigns the value 1 if trial i results in a success, and 0 if not. Then: $E(X_i) = 0 \cdot (1-p) + 1 \cdot p = p$. Let S denote the random variable which assigns the total number of successes: $S = \sum_{i=1}^{n} X_i$. Applying the sum theorem, we find that: $E(S) = \sum_{i=1}^{n} E(X_i) = \sum_{i=1}^{n} p = np$.

33. Assume that A will receive \$2 if a head comes up on the first toss (with probability 1/2); \$4 if on the second (with probability 1/4); \$8 if on the third (with probability 1/8); and so on. Suppose that A pays nothing to play the game. The expected winnings of A is then easily seen to be infinite:

$$\$2 \cdot \frac{1}{2} + \$4 \cdot \frac{1}{4} + \$8 \cdot \frac{1}{8} + \$16 \cdot \frac{1}{16} + \ldots = \$(1 + 1 + 1 + 1 + \ldots)$$

It follows that no matter how much you have to pay to play the game, you will still come out ahead, on the average. Yes, but would you pay, say, a million dollars to play?

35. Calculating the expected number of customers reached by the three options we have:

TV	100	90	80	70	60
Probability	0.15	0.20	0.30	0.25	0.10

$E(TV\#) = 100(0.15) + 90(0.20) + 80(0.30) + 70(0.25) + 60(0.10)$
$= 80.50$

Radio	75	70	60	50	40
Probability	0.25	0.25	0.25	0.20	0.05

$E(R\#) = 75(0.25) + 70(0.25) + 60(0.25) + 50(0.20) + 40(0.05)$
$= 63.25$

Magazines	50	40	30	20	10
Probability	0.50	0.30	0.10	0.06	0.04

$E(M\#) = 50(0.50) + 40(0.30) + 30(0.10) + 20(0.06) + 10(0.04)$
$= 41.60$

Decision: Go with the TV option which can be anticipated to reach 80,500 potential customers.

37. We need to compare the expected financial gain (Contributions minus Operating Cost) for the following tree options:

Door-To-Door Average Contribution: $15.70 per person		Mailing Average Contribution: $5.10 per person		Phone Average Contribution: $2.50 per person	
Number of contributors	Probability	Number of contributors	Probability	Number of contributors	Probability
3000	0.6	6000	0.4	11000	0.7
2500	0.3	5500	0.4	8000	0.2
2000	0.1	2500	0.2	7000	0.1
Operating cost: $17,000		Operating cost: $12,000		Operating cost: $6,000	

Expected number of contributors

Door-to-Door: $E(\text{Gain}) = \$15.70[3000(0.6) + 2500(0.3) + 2000(0.1)] - \$17,000$
$= \$43,175 - \$17,000 = \$26,175$

$$\text{Mailing:} \quad E(\text{Gain}) = \$5.10[\overbrace{6000(0.4)+5500(0.4)+2500(0.2)}^{\text{Expected number of contributors}}] - \$12,000$$
$$= \$26,010 - \$12,000 = \$14,010$$

$$\text{Phone:} \quad E(\text{Gain}) = \$2.50[\overbrace{11,000(0.7)+8000(0.2)+7000(0.1)}^{\text{Expected number of contributors}}] - \$6000$$
$$= \$25,000 - \$6000 = \$19,000$$

Decision: Go with the Door-to-Door option with an anticipated gain of $26,175.

39. We are given:

Ticket Requests	200	190	180	170	160	150	140
Probability	0.30	0.20	0.17	0.15	0.10	0.05	0.03

There are two options: Aircraft A or B. Each filled seat on either aircraft will bring in a revenue of $(350 - 75) = \$275$. The total operating expenses for aircraft A and B are given to be $40,000 and $45,000, respectively. While Aircraft B can accommodate up to 200 passengers, only 170 passengers can fly in aircraft A. Let's calculate the expected profit for the two aircraft, starting with B:

Aircraft A:

When it comes to aircraft A, it is important to note that all 170 of its seats will be filled whenever 170 **or more** individuals want to take the flight (if, say 184 people want to fly, then there may be 14 frustrated would-be travelers but all of the seats will still be filled). So, the probability that 170 seats will be filled is not the "0.15" in the above table, for that represents the probability that there are 170 requests for a seat. Rather, it is the probability at there are 170 **or more** requests; namely:

$$Pr(170 \text{ seats are filled}) = 0.30 + 0.20 + 0.17 + 0.15 = \textbf{0.82}$$

Bringing us to:

$$E(\text{Profit}) = \$275[170(\textbf{0.82}) + 160(0.10) + 150(0.05) + 140(0.03)] - \$40,000$$
$$= \$49,952.50 - \$40,000 = \$5,952.50$$

Aircraft B:

$$E(\text{Profit}) = \$275[\overbrace{200(0.30)+190(0.20)+180(0.17)+170(0.15)+160(0.10)+150(0.05)+140(0.03)}^{\text{expected number of seats filled}}] - \$45,000$$
$$= \$49,995 - \$45,000 = \$4995.00$$

Decision: Choose Aircraft A.

41. If you buy Collision: You will be out $250, and will still have to pay the $100 deductible if you are involved in an accident (probability 0.07). So:

$$E(\text{Additional cost for collision insurance}) = \$100(0.07) + \$250 = \$257$$

If you don't buy Collision then you will not pay $250 but can expect to have to pay $1200 with probability 0.07. So:

$$E(\text{Additional cost if you do not buy collision insurance}) = \$1200(0.07) = \$84$$

Decision: Don't buy the collision insurance.

43. In accordance with the following information:

Anticipated Number of Rental Requests	12	11	10	9	8	7	6
Probability	0.09	0.11	0.22	0.24	0.12	0.12	0.10

we need to determine the expected profit for the company in terms of the number of cars on hand.

For definiteness, let us focus on the **9 cars on hand** situation:

The best that can happen in this situation is that there are at least 9 rental requests (9 or 10 or 11 or 12), for then all 9 of the cars will be rented bringing in a $12 profit for each of the cars. The probability of this happening is: $Pr(12 \text{ or } 11 \text{ or } 10 \text{ or } 9) = \mathbf{0.09 + 0.11 + 0.22 + 0.24}$ [see **(*)** in the E(9-Profit) formula below]. There is, however, a probability of less than 9 requests (8 or 7 or 6). In each of these cases the requests can be accommodated giving rise to a $12 profit per rental, along with a $4 loss for each of the 9 cars that is not rented. In particular, if there are only 7 rental requests (with probability 0.12) then that will result in a profit of $\mathbf{7(\$12) - 2(\$4)}$ [see **(**)** in the E(9-Profit) formula below]. This reasoning process leads us to the following formula for the expected profit if the company has 9 cars on hand:

(*) profit if all 9 are rented   profit if 8 of the 9 are rented

$$E(\text{9-Profit}) = \$(9 \cdot 12)(\mathbf{0.09 + 0.11 + 0.22 + 0.24}) + \$(8 \cdot 12 - 4)(0.12)$$

$$+ \$(7 \cdot 12 - 2 \cdot 4)(0.12) + \$(6 \cdot 12 - 3 \cdot 4)(0.10) = \$97.44$$

(**) profit if 7 of the 9 are rented

If you understand the above 9-Profit expected value development, then you will see where the rest of the expected values are coming from:

$E(\text{12-Profit}) = \$(12 \cdot 12)(0.09) + \$(11 \cdot 12 - 4)(0.11) + \$(10 \cdot 12 - 2 \cdot 4)(0.22) + \$(9 \cdot 12 - 3 \cdot 4)(0.24)$
$\qquad + \$(8 \cdot 12 - 4 \cdot 4)(0.12) + \$(7 \cdot 12 - 5 \cdot 4)(0.12) + \$(6 \cdot 12 - 6 \cdot 4)(0.10) = \mathbf{\$96.80}$

$E(\text{11-Profit}) = \$(11 \cdot 12)(0.09 + 0.11) + \$(10 \cdot 12 - 4)(0.22) + \$(9 \cdot 12 - 2 \cdot 4)(0.24) + \$(8 \cdot 12 - 3 \cdot 4)(0.12)$
$\qquad + \$(7 \cdot 12 - 4 \cdot 4)(0.12) + \$(6 \cdot 12 - 5 \cdot 4)(0.10) = \mathbf{\$99.36}$

$E(\text{10-Profit}) = \$(10 \cdot 12)(0.09 + 0.11 + 0.22) + \$(9 \cdot 12 - 4)(0.24) + \$(8 \cdot 12 - 2 \cdot 4)(0.12)$
$\qquad + \$(7 \cdot 12 - 3 \cdot 4)(0.12) + \$(6 \cdot 12 - 4 \cdot 4)(0.10) = \mathbf{\$100.16}$

$E(\text{9-Profit}) = \$(9 \cdot 12)(0.09 + 0.11 + 0.22 + 0.24) + \$(8 \cdot 12 - 4)(0.12)$
$\qquad + \$(7 \cdot 12 - 2 \cdot 4)(0.12) + \$(6 \cdot 12 - 3 \cdot 4)(0.10) = \mathbf{\$97.44}$

$E(\text{8-Profit}) = \$(8 \cdot 12)(0.09 + 0.11 + 0.22 + 0.24 + 0.12) + \$(7 \cdot 12 - 4)(0.12)$
$\qquad + \$(6 \cdot 12 - 2 \cdot 4)(0.10) = \mathbf{\$91.48}$

$E(\text{7-Profit}) = \$(7 \cdot 12)(0.09 + 0.11 + 0.22 + 0.24 + 0.12 + 0.012) + \$(6 \cdot 12 - 4)(0.10) = \mathbf{\$82.40}$

$E(\text{6-Profit}) = \$6 \cdot 12 = \mathbf{\$12.00}$

Decision: The company should keep 10 cars on hand.

45. Assume you chose door 1 and that the host opens door 3. Before door 3 is opened, there is a $\frac{1}{3}$ probability that the \$100,000 is behind door 1, and a $\frac{2}{3}$ probability that it lies behind door 2 or door 3. The money is never moved, so after door 3 is opened, the probability that the money is behind door 1 remains at $\frac{1}{3}$. Now, the host knew the location of the money, and when he opens a door it will not be your door, nor will it be the "money door" which, with probability $\frac{2}{3}$, is not your door. That being the case there is a $\frac{2}{3}$ probability that the money door is not your door but is door 2. So **switch**, for the odds are two-to-one that the money is behind door 2.

---

§Review Exercises
Page 108

1. (a) $Pr(5 \text{ or } 7) = \frac{4+4}{52} = \frac{8}{52}$ (b) $Pr(5 \text{ of heart}) = \frac{1}{52}$

(c) $Pr(5 \text{ or } H) = Pr(5) + Pr(H) - Pr(5H) = \frac{4}{52} + \frac{13}{52} - \frac{1}{52} = \frac{16}{52}$

5 of spade or club or diomand $\searrow$ the 13 harts with the 5 of heart **removed** $\swarrow$

(d) $Pr(5 \text{ or } H \text{ but not } 5H) = \frac{3}{52} + \frac{12}{52} = \frac{15}{52}$

There are not 30 days in February $\swarrow$

2. (a) $Pr(\text{April } 15) = \frac{1}{365}$ (b) $Pr(30^{th}) = \frac{11}{365}$

There are 31 days in December $\searrow$

(c) $Pr(\text{December}) = \frac{31}{365}$ (d) $Pr(\text{Not in December}) = \frac{365 - 31}{365} = \frac{334}{365}$

2, 4, 6, 8, 10, 12, 14, 16   13, 14, 15, 16 $\searrow$ 4, 8, 12, 16 $\searrow$

3. (a) $Pr(\text{even}) = \frac{8}{15}$ (b) $Pr(> 12) = \frac{4}{15}$ (c) $Pr(\text{divisible by } 4) = \frac{4}{15}$

the smallsest number divisible by both 4 and 5 is 20 $\searrow$ 4, 5, 8, 10, 12, 15,16 $\searrow$

(d) $Pr(\text{divisible by } 4 \text{ and } 5) = \frac{0}{15} = 0$ (d) $Pr(\text{divisible by } 4 \text{ or } 5) = \frac{7}{15}$

see (d) $\downarrow$

(f) $Pr(\text{not divisible by } 4 \text{ or } 5) = 1 - \frac{7}{15} = \frac{8}{15}$

4. $Pr(< 3 \text{ letters}) = \frac{5}{11}$ $\leftarrow$ A, is, to, be, at $\leftarrow$ number of words in sentence

5. Odds in favour of a 7: 6 to 30   or 1 : 5

there are 6 sums of seven (see Figure 1.2, page 3)

The rest are not sevens

6. Odds against drawing an Ace: 48 to 4   or 12 : 1

there are 48 non-Aces

The rest are Aces

7. From the given information that the odds you will lose are 100 to 1 we have that the odds of winning are 1 to 100; which is to say that you will win out of 101 times. So $Pr(\text{win}) = \dfrac{1}{101}$.

8. Since $Pr(\text{rain}) = \dfrac{35}{100}$: "35 successes over 100 possibilities," there are "35 successes to 65 failures." Thus, the odds that it will rain are 35 to 65.

9. (a)
T: 26
S: 34
W: 64
TS: 8
TW: 5
SW: 2
TSW: 3

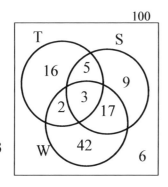

(b) $Pr(T\!\!\!/W) = \dfrac{16+5}{100} = \dfrac{21}{100}$

(c) $Pr(T \text{ or } W) = \dfrac{16+5+3+2+17+42}{100} = \dfrac{85}{100}$

(d) $Pr(S\!\!\!/W) = \dfrac{16+6}{100} = \dfrac{22}{100}$

10. The temptation is to exhibit both a Female circle and a Male circle in the Venn diagram. Don't try not to be so politically correct. You would not represent both a Campus circle and a non-Campus circle in the diagram, and just as everything outside the C-circle represent the students that do not live on campus, so then does everything outside the F-circle represent the students that are not female — the males:

(a) M stands for Math, not male

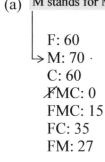

F: 60
M: 70 ·
C: 60
F̶MC: 0
FMC: 15
FC: 35
FM: 27

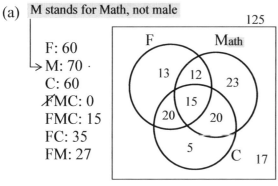

(b) $Pr(\cancel{\varnothing}) = \dfrac{125-60}{125} = \dfrac{65}{125}$

(c) $Pr(F\!\!\!/M) = \dfrac{13+20}{125} = \dfrac{33}{125}$

(d) $Pr(\cancel{F}\!M\cancel{C}) = \dfrac{17}{125}$

11. 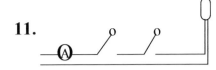 In order for the light to go on both switches must be good. We are dealing with independent events. So:

$$Pr(\text{On}) = (0.95)(0.95)$$

↑   ↑

first good and second good

12. 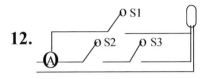 The light will go on if S1 is good OR if S1 is bad and both S2 and S3 are good:

$$Pr(\text{On}) = 0.95 + (0.05)(0.95)(0.95)$$

13.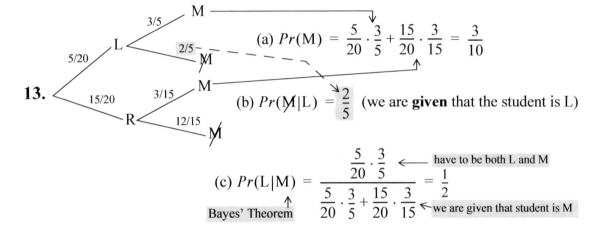

(a) $Pr(M) = \dfrac{5}{20} \cdot \dfrac{3}{5} + \dfrac{15}{20} \cdot \dfrac{3}{15} = \dfrac{3}{10}$

(b) $Pr(M|L) = \dfrac{2}{5}$ (we are **given** that the student is L)

(c) $Pr(L|M) = \dfrac{\dfrac{5}{20} \cdot \dfrac{3}{5}}{\dfrac{5}{20} \cdot \dfrac{3}{5} + \dfrac{15}{20} \cdot \dfrac{3}{15}} = \dfrac{1}{2}$

← have to be both L and M

↑ Bayes' Theorem

← we are given that student is M

14.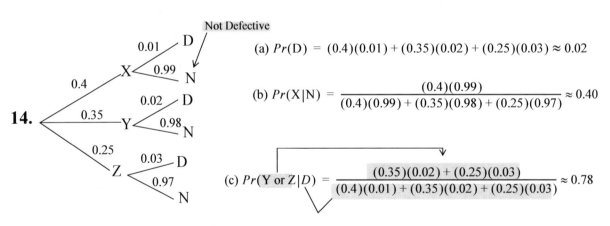

Not Defective

(a) $Pr(D) = (0.4)(0.01) + (0.35)(0.02) + (0.25)(0.03) \approx 0.02$

(b) $Pr(X|N) = \dfrac{(0.4)(0.99)}{(0.4)(0.99) + (0.35)(0.98) + (0.25)(0.97)} \approx 0.40$

(c) $Pr(Y \text{ or } Z|D) = \dfrac{(0.35)(0.02) + (0.25)(0.03)}{(0.4)(0.01) + (0.35)(0.02) + (0.25)(0.03)} \approx 0.78$

15.

Troll

0.1 La-La

T

0.9 Not

0.7

0.3

0.9 La-La

E

0.1 Not

(a) $Pr(\text{La-La}) = (0.7)(0.1) + (0.3)(0.9) = 0.34$

(b) $Pr(T|\text{La-La}) = \dfrac{(0.7)(0.1)}{(0.7)(0.1) + (0.3)(0.9)} \approx 0.21$

16. Choices followed by choices you multiply: $4 \cdot 2 \cdot 2 \cdot 7 \cdot 5 = 560$ available meals.

17. (a) $Pr(\text{no A}) \overset{\text{grab 5 of the 48 non-A}}{=} C(48, 5) = 1{,}712{,}304$ (b) $Pr(\text{no A or K}) \overset{\text{grab 5 of the 44 non-(A or K)}}{=} C(44, 5) = 1{,}086{,}008$

(c) $Pr(2 \text{ A and } 3 \text{ K}) \overset{\text{grab 2 of the 4 A AND 3 of the 4 K}}{=} C(4, 2)C(4, 3) = 24$

(d) $Pr(2 \text{ A and 2 or 3 K}) \overset{\text{2A, 2K and a non(A or K) OR 2A and 3K}}{=} C(4, 2)C(4, 2)C(44, 1) + C(4, 2)C(4, 3) = 1608$

18. First choose the order of the 13 types: 13! choices. Then order **each** of those types: 4! choices for each of the 13 types. From the Fundamental Counting Principle we conclude that there are: $13!(4!)^{13} = 5{,}457{,}011{,}114{,}581{,}223{,}249{,}490{,}739{,}200$ (a lot).

19.  (a) Put anyone in chair 1 (choice of 8). Choice of 1 for chair 2 (the other half of the couple). Any remaining person can sit in chair 3 (choice of 6). Choice of 1 for chair 4. Choice of 4 for chair 5 and choice of 1 for chair 6; choice of 2 for chair 7 and choice of 1 for chair 8. Choices followed by choices, you multiply:

Number of seating arrangements if no couiple is separtated $= 8 \cdot 1 \cdot 6 \cdot 1 \cdot 4 \cdot 1 \cdot 2 \cdot 1 = 384$

(b) You can start the 4-block of women-chairs in chair 1 through 5 (**choice of 5**) (can't start at 6 for there are only two chair to the right of 6). You still have a **choice of** 4! ways of sitting the women in the 4-block chosen, and another **choice of** 4! of sitting the men in the remaining chairs, for a grand total of $5 \cdot 4! \cdot 4! = 2280$ ways of seating the women next to each other.

(c) 4! ways one can order the men in thesse specified chairs

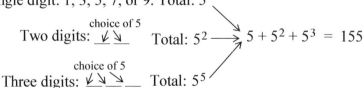

Total number of ways: $4! \cdot 4! = 576$

4! ways one can order the women in the other chairs

20. (a) Let's count the number of positive integers less than 1000 composed entirely of odd digits:

Single digit: 1, 3, 5, 7, or 9. Total: 5

Two digits: $\underset{\text{choice of 5}}{\underline{\quad\quad}}$ Total: $5^2 \longrightarrow 5 + 5^2 + 5^3 = 155$

Three digits: $\underset{\text{choice of 5}}{\underline{\quad\quad}}$ Total: 5^5

(b) Only even digits: Single digit: 2, 4, 6, or 8. Total: 4

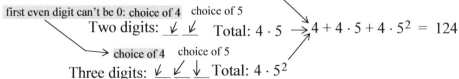

Two digits: $\underset{\text{first even digit can't be 0: choice of 4} \quad \text{choice of 5}}{\underline{\quad\quad}}$ Total: $4 \cdot 5 \longrightarrow 4 + 4 \cdot 5 + 4 \cdot 5^2 = 124$

Three digits: $\underset{\text{choice of 4} \quad \text{choice of 5}}{\underline{\quad\quad}}$ Total: $4 \cdot 5^2$

(c) At least two fives:

5 5	5 5 5	~~5~~ 5 5	5 ~~5~~ 5	5 5 ~~5~~
1	1	8	9	9
		↑ can't be 5 or 0	↑ can't be 5	↑ can't be 5

Total: $1 + 1 + 8 + 9 + 9 = 28$

(d) To have at least two zeros is to have **only** two zeros (must be a **positive** integer less than 1000). Those two zeros **must** occur like this: $\underline{\cancel{0}} \ \underline{0} \ \underline{0}$. Total: $\underset{\underset{\text{not } 0}{\uparrow}}{9} \cdot 1 \cdot 1 = 9$.

21. (a) Whatever sock you grab first is irrelevant. The second sock, however must match the first. Since there are 9 "second-socks" available, and since only one of the 9 match the first sock, the probability that you end up with a pair is $\frac{1}{9}$.

(b) One way: There are $C(10, 3)$ different ways of grabbing three socks from 10. To count the successes (ending up with a pair) we note that there is a **choice of 5** (the 5 different pairs), followed by a **choice of 8** (one of the other socks). Thus: $Pr(\text{pair}) = \frac{5 \cdot 8}{C(10, 3)} = \frac{1}{3}$.

Another way: $Pr(\text{pair}) = 1 - Pr(\text{all different}) = 1 - \frac{\overset{\text{grab 3 different types from the 5 types}}{\overset{\downarrow}{C(5, 3)} \cdot \overset{\text{choice of 2 for each of those 3 types}}{\overset{\downarrow}{2^3}}}}{C(10, 3)} = \frac{1}{3}$

(c) If you grab 6 socks you surely will have at least one pair. So, the probability is one.

22. Let's develop a sample space that reflects the fact that Big Boy (B) is two times as likely to win as Lucky (L) which is three times as likely to win as Sweet Lady (S):

$$\overset{\text{three more L's than S}}{\underset{\text{and twice as many B's as L's}}{S, L, L, L, B, B, B, B, B, B}}$$

Now let the race begin with each of the above 10 as likely to win as any other.

(a) $Pr(\text{S wins}) = \frac{1}{10}$.　　　(b) Odds against S winning: 9 to 1.

(c) $Pr(B) = \frac{6}{10}$.　　　(d) Odds in favor of B winning: 6 to 4.

23. (a)

$\begin{array}{|c|} \hline N_5 \ \ F_4 \\ P_3 \\ \hline \end{array}$ 　 $Pr(3P) = \frac{\overset{\text{3 of the 3 poetry}}{\overset{\downarrow}{C(3, 3)}}}{\underset{\underset{\text{grab 3 of the 12 books}}{\uparrow}}{C(12, 3)}} = \frac{1}{C(12, 3)} = \frac{1}{220}$

(b) $Pr(\text{one of each type}) = \frac{\overset{\text{one of the N and one of the F, and one of the P}}{C(5, 1)C(4, 1)C(3, 1)}}{C(12, 3)} = \frac{5 \cdot 4 \cdot 3}{C(12, 3)} = \frac{3}{11}$

(c) $Pr(\text{some P}) = 1 - Pr(\text{no P}) = 1 - \frac{C(3, 0)C(9, 3)}{C(12, 3)} = 1 - \frac{C(9, 3)}{C(12, 3)} = \frac{34}{55}$

24. (a) Pr(The 4 math books are first) $= \dfrac{\overset{\text{number of ways one can order the 4 math books}}{4! \cdot 8!}}{\underset{\text{number of ways one can order the 12 books}}{12!}} = \dfrac{1}{495}$

number of ways one can order the 4 math books   numer of ways one can order the 8 non-math books

number of ways one can order the 12 books

(b) The first math book can be in the first through the 9th position on the shelf — choice of 9

order the 4 math books

order the rest

$$Pr(\text{The 4 math books are together}) = \frac{9 \cdot 4! \cdot 8!}{12!} = \frac{1}{55}$$

(c) The number of ways you can order the **three type** of books   the number of ways you can then order the math books the art books, and the history books

$$Pr(\text{M together, A together, H together}) = \frac{3! \cdot 4! \cdot 5! \cdot 3!}{12!} = \frac{1}{4620}$$

25. (a)

$$\begin{array}{|cc|} \hline R_5 & W_4 \\ B_3 & Y_2 \\ \hline \end{array}$$

$\overset{4}{\nearrow}$

4 non-white

$Pr(\text{no W}) = \dfrac{C(10, 4)}{C(14, 4)} \approx 0.21$

grab 4 from 14

(b) $Pr(\text{all W}) = \dfrac{C(4, 4)}{C(14, 4)} = 0.001$

(c) $Pr(\text{at least one W}) = 1 - Pr(\text{no W})$

$$= 1 - \frac{C(10, 4)}{C(14, 4)} \approx 0.79$$

(d) B or R

$$Pr(\text{no W}|\text{no Y}) = \frac{C(8, 4)}{C(12, 4)} \approx 0.14$$

(e) 0W and 0Y   or   1W and 1Y   or   2W and 2Y

$$Pr(\text{same W as Y}) = \frac{C(8, 4) + C(4, 1)C(2, 1)C(8, 2) + C(4, 2)C(2, 2)}{C(14, 4)} \approx 0.30$$

(f) a success must consist of 1 R and 3 non-W

$$Pr(\text{no W}|\text{exactly 1 R}) = \frac{C(5, 1)C(5, 3)}{C(5, 1)C(9, 3)} = \frac{C(5, 3)}{C(9, 3)} \approx 0.012$$

in the sample space we have to have 1 R and 3 non-R

(g) $Pr(\text{all W}|\text{at least 3 W}) = \dfrac{C(4, 4)}{C(4, 3)C(10, 1) + C(4, 4)} = \dfrac{1}{C(4, 3)C(10, 1) + C(4, 4)} \approx 0.025$

in the sample space we have to have 3 or 4 W

26. (a)

$$\begin{array}{|cc|} \hline Q_2 & D_5 \\ N_3 & \\ \hline \end{array}$$

$\overset{2}{\nearrow}$

grab 2 of the 2 quarters

$Pr(50) = \dfrac{C(2, 2)}{C(10, 2)} = \dfrac{1}{45}$

grab 2 of the 10 coins

2 quarter and 1 nickel

(b) $Pr(30) = \dfrac{C(2, 1)C(3, 1)}{C(10, 2)} = \dfrac{2}{15}$

(c) 2 quarters or a quarter and a nickel or a quarter and a dime

$$\frac{C(2, 2) + C(2, 1)C(3, 1) + C(2, 1)C(5, 1)}{C(10, 2)} = \frac{17}{45}$$

(d) NOT 2 nickels

$$Pr(>10) = 1 - \frac{C(3, 2)}{C(10, 2)} = \frac{14}{15}$$

(e) $Pr(\geq 30 | \text{exactly one dime}) = \dfrac{\overset{\text{one dime and a quarter}}{C(5,1)\overset{\downarrow}{C}(2,1)}}{\underset{\underset{\text{one dime and one non-dime}}{\uparrow}}{C(5,1)C(5,1)}} = \dfrac{2}{5}$

(f) $Pr(<30 | \text{exactly one dime}) = \dfrac{\overset{\text{one dime and a nickel}}{C(5,1)\overset{\downarrow}{C}(3,1)}}{\underset{\underset{\text{one dime and one non-dime}}{\uparrow}}{C(5,1)C(5,1)}} = \dfrac{3}{5}$

27. (a) $\boxed{\begin{array}{l} Q_2 \\ D_5 \\ N_3 \end{array}}\overset{\nearrow 4}{}$ $\underset{\text{grab 4 of the 10 coins} \rightarrow}{Pr(50)} = \dfrac{\overset{\text{a quarter and 2 dimes and 1 nickel}}{C(2,1)\overset{\downarrow}{C}(5,2)\overset{\downarrow}{C}(3,1)}}{C(10,4)} = \dfrac{2}{7}$

(b) $Pr(60) = \dfrac{\overset{\text{2 quarter and 2 nickels}}{C(2,2)\overset{\downarrow}{C}(3,2)}}{C(10,2)} = \dfrac{1}{70}$ (c) $Pr(55 | \text{no nickel}) = \dfrac{\overset{\text{one quarter and 3 dimes}}{C(2,1)\overset{\downarrow}{C}(5,3)}}{\underset{\underset{\text{grab 4 of the non-nickels}}{\uparrow}}{C(7,4)}} = \dfrac{4}{7}$

28. (a) $\boxed{\begin{array}{l} \boxed{\text{B J M}} \\ \boxed{\text{7 Others}} \end{array}}\overset{\nearrow 3 \;\text{In addition to B and J you need one more (can be M)}}{}$ $\underset{\text{grab 3 of the 10 people} \nearrow}{Pr(\text{B and J})} = \dfrac{C(8,\overset{\downarrow}{1})}{C(10,3)} = \dfrac{1}{15}$

(b) $Pr(\text{B or J but not both}) = \dfrac{\overset{\text{choice of 2 (B or J) and one of the 8 other}}{\searrow 2\cdot 8 \swarrow}}{C(10,3)} = \dfrac{2}{15}$

(c) $Pr(\text{B or J, possibly both}) = \dfrac{\overset{\text{B or J but not both or both B and J and one of the remaining 8}}{2\cdot 8 \overset{\downarrow}{+} 1\cdot 8}}{C(10,3)} = \dfrac{17}{120}$

(d) $Pr(\text{B, J, and M}) = \dfrac{1}{C(10,3)} = \dfrac{1}{120}$ (e) $Pr(\text{ecactly two}) = \dfrac{\overset{\text{2 or the 3 and 1 of the other 7}}{C(3,2)\overset{\downarrow}{C}(7,1)}}{C(10,3)} = \dfrac{7}{40}$

29. There are 900 integer between 100 and 999 (inclusive): $\overset{\text{choice of 9 choice of 10}}{\underset{\downarrow\;\;\downarrow\;\;\downarrow}{}}$. So:

(a) $Pr(\text{3 digits the same}) = \dfrac{\overset{\text{choice of 9 for first digit}}{\overset{\downarrow}{9}\cdot 1\cdot 1}}{900} = \dfrac{1}{10}$ (b) $Pr(\text{no 0}) = \dfrac{9\cdot 9\cdot 9}{900} = \dfrac{81}{100}$

(c) $Pr(\text{odd}) = \dfrac{\overset{\text{last digit must be odd}}{9\cdot 10\cdot \overset{\downarrow}{5}}}{900} = \dfrac{1}{2}$ (d) $Pr(\text{no repeated digit}) = \dfrac{\overset{\text{can't be the first can't be the first two}}{9\cdot \overset{\downarrow}{9}\cdot \overset{\downarrow}{8}}}{900} = \dfrac{18}{25}$

30.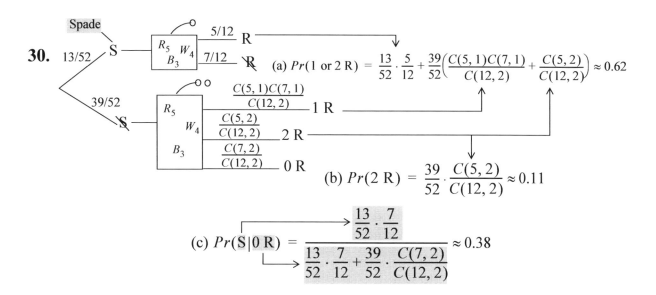

$$(a)\ Pr(1\ or\ 2\ R) = \frac{13}{52}\cdot\frac{5}{12} + \frac{39}{52}\left(\frac{C(5,1)C(7,1)}{C(12,2)} + \frac{C(5,2)}{C(12,2)}\right) \approx 0.62$$

$$(b)\ Pr(2\ R) = \frac{39}{52}\cdot\frac{C(5,2)}{C(12,2)} \approx 0.11$$

$$(c)\ Pr(S|0\ R) = \frac{\frac{13}{52}\cdot\frac{7}{12}}{\frac{13}{52}\cdot\frac{7}{12} + \frac{39}{52}\cdot\frac{C(7,2)}{C(12,2)}} \approx 0.38$$

31.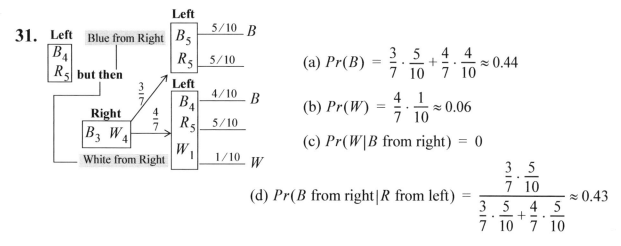

$$(a)\ Pr(B) = \frac{3}{7}\cdot\frac{5}{10} + \frac{4}{7}\cdot\frac{4}{10} \approx 0.44$$

$$(b)\ Pr(W) = \frac{4}{7}\cdot\frac{1}{10} \approx 0.06$$

$(c)\ Pr(W|B\ \text{from right}) = 0$

$$(d)\ Pr(B\ \text{from right}|R\ \text{from left}) = \frac{\frac{3}{7}\cdot\frac{5}{10}}{\frac{3}{7}\cdot\frac{5}{10} + \frac{4}{7}\cdot\frac{5}{10}} \approx 0.43$$

32. (a) There are 4! ways individual X can arrange his four cards, only one of which matches the order of individual Y. So: $Pr(\text{all match}) = \frac{1}{4!} = \frac{1}{24}$.

(b) If 3 of X's cards match 3 of Y's cards, then the fourth card of X **has to match** the fourth card of Y. So: $Pr(\text{exactly 3 mathces}) = 0$.

(c) Let's look at a success: X matches exactly 2 of Y's cards: Y: A K J Q / X: A J K Q. We decided to have X match the Aces and Queen, but it could have been any 2 of the four type of cards — **a choice of** $C(4,2)$. One the Ace and Queen were matched, the Jack and King **had** to be mis-matched (a **choice of 1**). So: $Pr(\text{exactly mathces}) = \frac{C(4,2)\cdot 1}{4!} = \frac{1}{4}$.

(d) Chose the one card that matches: a **choice of 4** (for example: Y: A K J Q / X: K Q J A). Now looking at our example, we see that under the A we had a **choice of 2** (not A nor J). The rest was forced on us (had to put Q under the K and the A under the Q). So: $Pr(1\ \text{match only}) = \frac{4\cdot 2}{4!} = \frac{1}{3}$.

(e) One way: Let us assume a specific order for Y, say A K Q J. If there is to be no match, then under A we have a **choice of 3**. For definiteness, let's place K under A and go on from there: Y: A K J   Q, Systematically listing all subsequent possibilities we have:

K

Y: A K J Q

choice of 3 →K A   Q J
              K J   A K → choice of 3   . So: $Pr(\text{No match}) = \dfrac{3 \cdot 3}{4!} = \dfrac{3}{8}$.
              K Q   A J

Another way:

$$Pr(\text{No match}) = 1 - [Pr(4 \text{ match}) + Pr(3 \text{ match}) + Pr(2 \text{ match}) + Pr(\text{ match})]$$

$$= 1 - \left(\frac{1}{24} + 0 + \frac{1}{4} + \frac{1}{3}\right) = \frac{3}{8}$$

from (a) (b) (c) (d)

33. (a)

n	p	r
5	$\frac{2}{3}$	3

$Pr(\text{exactly 3 H}) = C(5,3)\left(\frac{2}{3}\right)^3\left(\frac{1}{3}\right)^2 \approx 0.33$

Since Head is twice as likely as a Tail

(b) $Pr(4 \text{ or } 5 \text{ H}) = C(5,4)\left(\frac{2}{3}\right)^4\left(\frac{1}{3}\right)^1 + C(5,5)\left(\frac{2}{3}\right)^5\left(\frac{1}{3}\right)^0 = C(5,4)\left(\frac{2}{3}\right)^4\left(\frac{1}{3}\right)^1 + \left(\frac{2}{3}\right)^5 \approx 0.46$

Not no H

(c) $Pr(\text{at least 1 H}) = 1 - Pr(\text{no H}) = 1 - C(5,0)\left(\frac{2}{3}\right)^0\left(\frac{1}{3}\right)^5 = 1 - \left(\frac{1}{3}\right)^5 \approx 0.996$

3 H and 2 T   or   4 H and 1 T   or 5 H and 0 T

(d) $Pr(\text{more H than T}) = C(5,3)\left(\frac{2}{3}\right)^3\left(\frac{1}{3}\right)^2 + C(5,4)\left(\frac{2}{3}\right)^4\left(\frac{1}{3}\right)^1 + C(5,5)\left(\frac{2}{3}\right)^5\left(\frac{1}{3}\right)^0 \approx 0.79$

34. (a)

n	p	r
10	$\frac{4}{52}$	4

$Pr(\text{exactly 4 draw an A}) = C(10,4)\left(\frac{4}{52}\right)^4\left(\frac{48}{52}\right)^6 \approx 0.0045$

Not no A

(b) $Pr(\text{at least one draws an A}) = 1 - C(10,0)\left(\frac{4}{52}\right)^0\left(\frac{48}{52}\right)^{10} \approx 0.55$

(c) $Pr(<3) = C(10,0)\left(\frac{4}{52}\right)^0\left(\frac{48}{52}\right)^{10} + C(10,1)\left(\frac{4}{52}\right)^1\left(\frac{48}{52}\right)^9 + C(10,2)\left(\frac{4}{52}\right)^2\left(\frac{48}{52}\right)^8 \approx 0.93$

35. (a)

n	p	r
100	0.1	10

$Pr(\text{exactly 10 exposed}) = C(100, 10)(0.1)^{10}(0.9)^{90} \approx 0.013$

(b) Probability of coming down with the flew is the probability that the person has the flew times the probability that the shot is not effective: $p = (0.1)(0.2) = 0.02$. So:

n	p	r
100	0.02	10

$Pr(\text{exactly 10 get the flu}) = C(100, 10)(0.02)^{10}(0.98)^{99} \approx 0.000029$

(c) $Pr(\text{at least one will get the flu}) = 1 - \overset{\overset{\text{Not none}}{\downarrow}}{C}(100, 0)(0.02)^{0}(0.98)^{100} \approx 1 - (0.98)^{100} \approx 0.87$

(d) $Pr(0 \text{ or } 1) = C(100, 0)(0.02)^{0}(0.98)^{100} + C(100, 1)(0.02)^{1}(0.98)^{99} \approx 0.40$

36.

$$\begin{array}{|l|}\hline D_5 \xrightarrow{7} \\ G_{95} \\\hline\end{array}$$

5 Defective and 95 Good

(a) $Pr(\text{exactly 3 D}) = \dfrac{C(5, 3)C(95, 7)}{C(100, 7)} \approx 0.002$

(b) $Pr(\text{at least 1 D}) = 1 - Pr(\text{no D}) = 1 - \dfrac{C(5, 0)C(95, 7)}{C(100, 7)} \approx 0.31$

(c) $Pr(\text{at most 6}) = 1 - \overset{\overset{\text{Not 7}}{\downarrow}}{Pr}(7) = 1 - \dfrac{C(7, 7)C(95, 0)}{C(100, 7)} = 1 - \dfrac{1}{C(100, 7)} \approx 0.99999999994$

37. (a) Here is the probability of drawing two clubs: $p = \dfrac{13}{52} \cdot \dfrac{12}{51} = \dfrac{1}{17}$. So:

n	p	r
5	$\dfrac{1}{17}$	4

$Pr(\text{exactly 4 draw 2 clubs exposed}) = C(5, 4)\left(\dfrac{1}{17}\right)^{4}\left(\dfrac{16}{17}\right)^{1} \approx 0.000056$

(b) $Pr(\text{at least one draws 2 clubs}) = 1 - \overset{\overset{\text{Not none}}{\downarrow}}{C}(5, 0)\left(\dfrac{1}{17}\right)^{0}\left(\dfrac{16}{17}\right)^{5} \approx 0.26$

(c) $Pr(0 \text{ or } 1 \text{ will draw 2 clubs}) = C(5, 0)\left(\dfrac{1}{17}\right)^{0}\left(\dfrac{16}{17}\right)^{5} + C(5, 1)\left(\dfrac{1}{17}\right)^{1}\left(\dfrac{16}{17}\right)^{4} \approx 0.97$

38. (a) Turning to Theorem 1.12, page 80 we have:

$Pr(\text{exactly 4 of the 10 draw 3 blue marbles}) = C(10, 4)p^{4}(1-p)^{6}$

$$\begin{array}{|l|}\hline R_3 \xrightarrow{3} B_5 \\ W_4 \\\hline\end{array}$$ where $p = Pr(3 \text{ blue}) = \dfrac{C(5, 3)}{C(12, 3)} = \dfrac{1}{22}$

So: $Pr(\text{exactly 4 of the 10 draw 3 blue marbles}) = C(10, 4)\left(\dfrac{1}{22}\right)^{4}\left(\dfrac{21}{22}\right)^{6} \approx 0.00068$

(b) $Pr(\text{at east 1 draw 2 blue}) = 1 - Pr(\text{none draw 2}) = 1 - C(10, 0)\left(\dfrac{1}{22}\right)^{0}\left(\dfrac{21}{22}\right)^{10} \approx 0.372$

(c) $Pr(\text{exactly 2 of the 10 draw RWB}) = C(10, 2)p^2(1-p)^8$

$\boxed{\begin{array}{c} R_3 \quad B_5 \\ W_4 \end{array}}$ $\nearrow 3$ where $p = Pr(RWB) = \dfrac{C(3,1)C(4,1)C(5,1)}{C(12,3)} = \dfrac{3}{11}$

So: $Pr(\text{exactly 2 of the 10 draw RWB}) = C(10,2)\left(\dfrac{3}{11}\right)^2\left(\dfrac{8}{11}\right)^8 \approx 0.026$

(d) $Pr(\text{at least 9 of the 10 draw at least 2 W}) = C(10,9)p^9(1-p)^1 + C(10,10)p^{10}(1-p)^0$

where $p = \dfrac{C(4,2)C(8,1)+C(4,3)}{C(12,3)} \approx \dfrac{13}{55}$

So: $Pr(\text{at least 9 of the 10 draw at least 2 W}) = C(10,9)\left(\dfrac{13}{55}\right)^9\left(\dfrac{42}{15}\right)^1 + \left(\dfrac{13}{55}\right)^{10} \approx 0.000065$

39. $\boxed{\begin{array}{c} R_3 \quad B_5 \\ W_4 \end{array}}$ $\nearrow 5$

(a) $Pr(3R) = \dfrac{C(3,3)C(9,2)}{C(12,5)} = \dfrac{1}{22}$

(b) $Pr(2R \text{ or } 3R) = \dfrac{C(3,2)C(9,3)+C(3,3)C(9,2)}{C(12,5)} = \dfrac{4}{11}$

(c) $Pr(<3R) = 1 - Pr(3R) = 1 - \dfrac{1}{22} = \dfrac{21}{22}$
$\underset{(a)}{\uparrow}$

(d) $Pr(3R|1W) = \dfrac{C(4,1)C(3,3)C(5,1)}{C(4,1)C(11,4)} = \dfrac{1}{66}$

40. $\boxed{\begin{array}{c} G_{95} \\ D_5 \end{array}}$ $\nearrow 2$
6 boxes

$Pr(0D) = \dfrac{C(95,2)}{C(100,2)} = \dfrac{893}{990}$

$Pr(2D) = \dfrac{C(5,2)}{C(100,2)} = \dfrac{1}{495}$ $Pr(\text{not } 2D) = 1 - \dfrac{1}{495} = \dfrac{494}{495}$

(a)

n	p	r
6	$\dfrac{893}{990}$	6

$Pr(6 \text{ have } 0D) = C(6,6)(\tfrac{893}{990})^6(1-\tfrac{893}{990})^0 = (\tfrac{893}{990})^6 = 0.54$

(b)

n	p	r
6	$\dfrac{494}{495}$	6

$Pr(6 \text{ have at most } 1D) = C(6,6)(\tfrac{494}{495})^6(1-\tfrac{494}{495})^0 = (\tfrac{494}{495})^6 \approx 0.988$

(c)

n	p	r
6	$\dfrac{1}{495}$	2

$Pr(2 \text{ have } 2D) = C(6,2)(\tfrac{1}{495})^2(1-\tfrac{1}{495})^4 = 0.000061$

41. Let's focus on some of the probabilities resting on the branches of the tree below. Specifically, what is the probability that William passes (exactly) 2 or the 4 courses? Ask Bernoulli:

$$\begin{array}{ccc} n & p & r \\ 4 & 0.6 & 2 \end{array} \longrightarrow Pr(\mathbf{2}) = C(4,2)(0.6)^2(0.4)^2$$

And, if he passes **2**, then a dice is rolled and he will go to Disney land if a 6 is rolled (probability: $\frac{1}{6}$); see tree below.

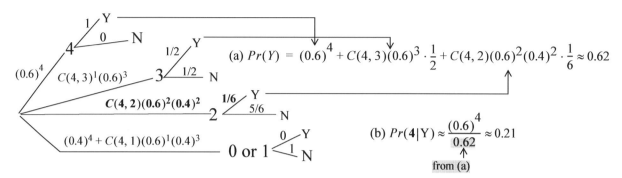

(a) $Pr(Y) = (0.6)^4 + C(4,3)(0.6)^3 \cdot \frac{1}{2} + C(4,2)(0.6)^2(0.4)^2 \cdot \frac{1}{6} \approx 0.62$

(b) $Pr(4|Y) \approx \frac{(0.6)^4}{0.62} \approx 0.21$

from (a)

42. $E(\text{Wining}) = \$2\left(\frac{1}{6}\right) + \$4\left(\frac{1}{6}\right) + \$6\left(\frac{1}{6}\right) - \$1\left(\frac{1}{6}\right) - \$3\left(\frac{1}{6}\right) - \$5\left(\frac{1}{6}\right)$

$$= \$(2 + 4 + 6 - 1 - 3 - 5)\left(\frac{1}{6}\right) = \$0.50$$

43. $E(\text{Commission}) = \$100(0.03) + \$75(0.05) + \$50(0.04) = \$8.75$

44. $E(\text{\# of Aces}) = 0 \cdot \dfrac{C(4,0)C(48,3)}{C(52,3)} + 1 \cdot \dfrac{C(4,1)C(48,2)}{C(52,3)} + 2 \cdot \dfrac{C(4,2)C(48,1)}{C(52,3)}$

$$+ 3 \cdot \dfrac{C(4,2)C(48,0)}{C(52,3)} = 0.23$$

45. **HHHH**
HHHT HTHH HTHH **THHH**
The rest have less than 3 Heads.
So there are **3 successes**.
Number of possibilities: $2 \cdot 2 \cdot 2 \cdot 2 = 2^4 = \mathbf{16}$
choice of 2 (H or T) four times

$E(\text{Winning}) = \$10\left(\frac{3}{16}\right) - \$2\left(\frac{13}{16}\right) = \$0.25$

Play: You can expect to win a quarter per game.

46. Option A: $E(\text{Profit}) = (-5000)(0.75) + (20{,}000)(0.25) = \$1250.00$

Option B: $E(\text{Profit}) = -5000(0.05) - 2500(0.55) + 0(0.10) + 5000(0.30) = -\$125.00$

Option C: $E(\text{Profit}) = \$500.00$

Conclusion: Choose option A.

§2.1. Preliminaries
Page 127

1. $-2^2(2+3)^2 = -4 \cdot 5^2 = -100$

Note: -2^2 is **NOT** the same as $(-2)^2 = (-2)(-2) = 4$; but, rather: $-2^2 = -4$.

3. $\dfrac{(2+3)^2}{3^2} = \dfrac{5^2}{3^2} = \dfrac{25}{9}$

Note: $(2+3)^2$ is **NOT** equal to $2^2 + 3^2$: $(2+3)^2 = 5^2 = 25$, while $2^2 + 3^3 = 4 + 9 = 13$.

5. $\dfrac{\left(\frac{1}{2} + \frac{1}{4}\right)^2}{2^3} = \dfrac{\left(\frac{2}{4} + \frac{1}{4}\right)^2}{8} = \dfrac{\left(\frac{3}{4}\right)^2}{8} = \dfrac{\frac{9}{16}}{\frac{8}{1}} \underset{\text{invert and multiply}}{\uparrow} \dfrac{9}{16} \cdot \dfrac{1}{8} = \dfrac{9}{128}$

7. $\dfrac{2^{-1} + 2}{9 \cdot 3^{-1}} = \dfrac{\frac{1}{2} + 2}{9 \cdot \frac{1}{3}} = \dfrac{\frac{1}{2} + \frac{4}{2}}{3} = \dfrac{\frac{5}{2}}{3} = \dfrac{5}{2} \cdot \dfrac{1}{3} = \dfrac{5}{6}$

9. One approach:

$$\dfrac{3^{-2} - 2^{-3}}{-3^2} = \dfrac{\frac{1}{3^2} - \frac{1}{2^3}}{-\frac{1}{3^2}} = \dfrac{\frac{1}{9} - \frac{1}{8}}{-\frac{1}{9}} = -\dfrac{\frac{1 \cdot 8}{9 \cdot 8} - \frac{1 \cdot 9}{8 \cdot 9}}{\frac{1}{9}} = -\dfrac{\frac{8-9}{9 \cdot 8}}{\frac{1}{9}} = -\dfrac{-1}{9 \cdot 8} \cdot \dfrac{9}{1} = \dfrac{-1}{8} = \dfrac{1}{8}$$

(see below)

Another approach: $\dfrac{3^{-2} - 2^{-3}}{-3^2} = \dfrac{\frac{1}{3^2} - \frac{1}{2^3}}{-\frac{1}{3^2}} = \dfrac{\frac{1}{9} - \frac{1}{8}}{-\frac{1}{9}} = -\dfrac{\left(\frac{1}{9} - \frac{1}{8}\right) \cdot 9}{\frac{1}{9} \cdot 9} = -\left(1 - \dfrac{9}{8}\right) = -\left(-\dfrac{1}{8}\right) = \dfrac{1}{8}.$

11. One approach: $\dfrac{(ab)^2}{a^3(-b)^3} = \dfrac{a^2 b^2}{-a^3 b^3} \uparrow = -\dfrac{\cancel{a^2}\cancel{b^2}}{a\cancel{a^2}\cancel{b^2}b} = -\dfrac{1}{ab}$

(see below)

The minus sign in $(-b)^3$ survives, since the expression involves an odd power: $(-b)^3 = -b^3$.
Were it raised to an even power, it would not survive; for example: $(-b)^4 = b^4$.

Another approach: $\dfrac{(ab)^2}{a^3(-b)^3} = -\dfrac{a^2 b^2}{a^3 b^3} \uparrow = -a^{2-3}b^{2-3} = -a^{-1}b^{-1} \uparrow = -\dfrac{1}{ab}$

Theorem 2.1(ii)   Definition 2.2

13. Having justified the fact that you can bring any term from one side of an equation to the other as long as you change its sign (see page 117), we have:

$$2x + 8 = 7x - 5$$
$$2x - 7x = -8 - 5$$
$$-5x = -13$$
$$x = \frac{13}{5}$$

15. $$-3x + 7 = 4x - 5 + \frac{x}{2}$$

$$-3x - 4x - \frac{x}{2} = -5 - 7$$

$$-7x - \frac{x}{2} = -12 \longrightarrow \text{or: multiply both sides by 2:} \quad -14x - x = -24$$

$$-\frac{14x}{2} - \frac{x}{2} = -12 \qquad\qquad -15x = -24$$

$$-\frac{15x}{2} = -12 \qquad\qquad x = \frac{24}{15} = \frac{8}{5}$$

$$-15x = -24$$

$$x = \frac{24}{15} = \frac{8}{5} \leftarrow \text{see below}$$

17. The best approach here is to clear denominators by multiplying both side of the given equation by 6:

$$\frac{x-2}{3} = 2(-x+1)$$

$$\overset{2}{6}\left(\frac{x-2}{3}\right) = 6[2(-x+1)]$$

$$2(x-2) = 12(-x+1)$$
$$2x - 4 = -12x + 12$$
$$2x + 12x = 12 + 4$$
$$14x = 16$$
$$x = \frac{16}{14} = \frac{8}{7}$$

19. $3x - 5 > 2x + 4$

$3x - 2x > 5 + 4$

$x > 9$

or: $(9, \infty)$

see page 122

21. $\dfrac{x}{2} - 3x \geq 4x + 5$

$2\left(\dfrac{x}{2} - 3x\right) \geq 2(4x + 5)$

$x - 6x \geq 8x + 10$

$x - 6x - 8x \geq 10$

$-13x \geq 10$

$x \leq -\dfrac{10}{13}$

or: $\left(-\infty, -\dfrac{10}{13}\right]$

see page 122

23. $\dfrac{x + 2}{4} + x - 1 < x + 2$

$4\left(\dfrac{x + 2}{4} + x - 1\right) < 4(x + 2)$

$x + 2 + 4x - 4 < 4x + 8$

$x + 4x - 4x < 8 - 2 + 4$

$x < 10$

or: $(-\infty, 10)$

see page 122

25. For $(2, 5), (3, 7)$:

$m = \dfrac{5 - 7}{2 - 3} = \dfrac{-2}{-1} = 2$

or, if you prefer:

$m = \dfrac{7 - 5}{3 - 2} = \dfrac{2}{1} = 2$

27. For $(2, -5), (3, 7)$:

$m = \dfrac{-5 - 7}{2 - 3} = \dfrac{-12}{-1} = 12$

or, if you prefer:

$m = \dfrac{7 - (-5)}{3 - 2} = \dfrac{12}{1} = 12$

29. For $(3, 6), (7, 6)$:

$m = \dfrac{6 - 6}{7 - 3} = \dfrac{0}{4} = 0$

or, if you prefer:

$m = \dfrac{6 - 6}{3 - 7} = \dfrac{0}{-4} = 0$

31.

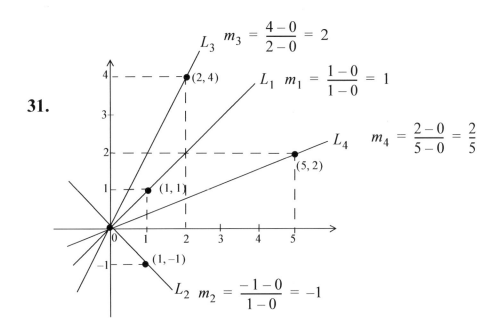

$L_3 \quad m_3 = \dfrac{4 - 0}{2 - 0} = 2$

$L_1 \quad m_1 = \dfrac{1 - 0}{1 - 0} = 1$

$m_4 = \dfrac{2 - 0}{5 - 0} = \dfrac{2}{5}$

$L_2 \quad m_2 = \dfrac{-1 - 0}{1 - 0} = -1$

33. For a point (x_0, y_0) to lie on the (graph of the) line $y = 4x + 3$, y_0 must equal $4x_0 + 3$. Lets check that requirement for the given points $(7, 1), (1, 7), (-2, 5), (-2, -5)$:

For $(7, 1)$: Is $1 = 4 \cdot 7 + 3$? No, so $(7, 1)$ does not lie on the line.

For $(1, 7)$: Is $7 = 4 \cdot 1 + 3$? Yes, so $(1, 7)$ does lie on the line.

For $(-2, 5)$: Is $5 = 4(-2) + 3$? No, so $(-2, 5)$ does not lie on the line.

For $(-2, -5)$: Is $-5 = 4(-2) + 3$? Yes, so $(-2 - 5)$ lies on the line.

35. Since there are infinitely many lines with y-intercept 9, there is no unique answer. Just let m be any two different numbers in the equation $y = mx + 9$, (For example: $y = 17x + 9$ and $y = -3x + 9$).

37. Every equation of a line of slope 0 is of the form $y = mx + b = 0x + b$. The line of slope 0 that we are looking for has to contain the point: $(1, 2)$; leading us to: $2 = 0 \cdot 1 + b$, or $b = 2$. So the equation of line of slope 0 passing through $(1, 2)$ is $y = 0x + 2$ or simply $y = 2$ (see graph on the right).

39. We need to find m and b in the equation $y = mx + b$. Using the two given points $(2, 1), (1, 2)$, we determine m: $m = \dfrac{2-1}{1-2} = \dfrac{1}{-1} = -1$.

At this point we know that our line is of the form: $y = -1 \cdot x + b$. You can choose either of the two given points and determine the value of b; let's go with $(2, 1)$:

$(2, 1)$

$1 = -1 \cdot 2 + b \Rightarrow b = 3$ Equation: $y = -x + 3$

41. Using the two given points $(-1, 3), (5, 3)$, we find that $m = \dfrac{3-3}{5-(-1)} = \dfrac{0}{6} = 0$; bringing us

to: $y = 0 \cdot x + b$. Since $(5, 3)$ is on the line (could have chosen the other point):

$(5, 3)$

$y = 0 \cdot x + b$: $3 = 0 \cdot 5 + b \Rightarrow b = 3$ So: $y = 0 \cdot x + 3$

Or simply: $y = 3$

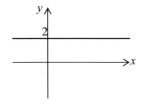

43. Since the slope is **3**: $y = 3x + b$. Since $(1, 3)$ is on the line:

$$(1, 3)$$
$$3 = 3 \cdot 1 + b \Rightarrow b = 3 - 3 = 0 \qquad \text{Equation: } y = 3x$$

45. Since the slope is **3**: $y = 3x + b$. Since $(3, 5)$ is on the line:

$$(3, 5)$$
$$5 = 3 \cdot 3 + b \Rightarrow b = 5 - 9 = -4 \qquad \text{Equation: } y = 3x - 4$$

47. Using the two given points $(2, -4)$, $(-3, 5)$, we find that $m = \dfrac{5 - (-4)}{-3 - 2} = -\dfrac{9}{5}$; bringing us to:

$y = -\dfrac{9}{5} \cdot x + b$. Since $(2, -4)$ is on the line (could have chosen the other point):

$$(2, -4)$$
$$-4 = -\frac{9}{5} \cdot 2 + b \Rightarrow b = -4 + \frac{18}{5} = -\frac{2}{5} \qquad \text{Equation: } y = -\frac{9}{5}x - \frac{2}{5}$$

49. The given line $y = 3x + 4$ has slope **3**. Being parallel to that line, "our line" also has slope 3 and is of the form $y = 3x + b$. Since our line contains the point $(3, -4)$: $-4 = 3 \cdot 3 + b \Rightarrow b = -13$. Answer: $y = 3x - 13$.

51. The given line $y = -2x + 1$ has slope **–2**. Being perpendicular to that line, "our line" has slope $-\dfrac{1}{-2} = \dfrac{1}{2}$ and is of the form $y = \dfrac{1}{2}x + b$. Since it is to contain the point $(2, 1)$:

$1 = \dfrac{1}{2} \cdot 2 + b \Rightarrow b = 0$. Answer: $y = \dfrac{1}{2}x$.

53. (a) Using the given slope $m = 3$ and the given point $(-2, 1)$ we arrive at the **point-slope** form:
$y - 1 = 3[x - (-2)] \Rightarrow y - 1 = 3(x + 2)$.

(b) Starting with $y = \underset{\text{given slope}}{3x} + b$ we use the given point $(-2, 1)$ to find b:

$1 = 3(-2) + b \Rightarrow b = 7$ bringing us to the slope-intercept form: $y = 3x + 7$.

(c) Starting with the point-slope form of (a): $y - 1 = 3(x + 2)$, we solve for y in terms of x:

$$\text{point-slope:} \quad y - 1 = 3(x + 2)$$
$$y - 1 = 3x + 6$$
$$\text{slope-intercept:} \quad y = 3x + 7$$

55. Applying the difference of two square formula we have: $9x^2 - 4 = (3x + 2)(3x - 2)$.

57. Applying the difference of two square formula we have: $4x^2 - 25 = (2x + 5)(2x - 5)$.

59. $-4x^2 + 1 = 1 - 4x^2 = (1 + 2x)(1 - 2x)$ or: $-4x^2 + 1 = -(4x^2 - 1) = -(2x + 1)(2x + 1)$

Note: $\overset{\text{here}}{-(2x+1)}(2x-1) = (-2x-1)(2x+1)$ or $\overset{\text{here}}{-(2x+1)(2x-1)} = (2x+1)(1-2x)$

but not in both of the factors

61. Applying the difference of two square formula we have:

$$[a^2 - b^2] = (a+b)(a-b)$$
$$[(\sqrt{2}x)^2 - (\sqrt{5})^2] = (\sqrt{2}x + \sqrt{5})(\sqrt{2}x - \sqrt{5})$$

63. $x^2 + 7x + 12 = (x+4)(x+3)$
$3x + 4x$

65. $x^2 - 7x + 12 = (x-4)(x-3)$
$-3x - 4x$

67. $6x^2 + 7x + 2 = (2x+1)(3x+2)$
$4x + 3x$

69. $6x^2 - 7x - 5 = (2x+1)(3x-5)$
$-10x + 3x$

71. $6x^3 + 31x^2 + 40x = x(6x^2 + 31x + 40) = x(2x+5)(3x+8)$
pull out common factor $16x + 15x$

73.
$$2x^2 + 9x - 35 = 0$$
$$(2x-5)(x+7) = 0$$
$$2x - 5 = 0 \qquad x + 7 = 0$$
$$x = \frac{5}{2} \qquad x = -7$$

75.
$$25x^2 - 16 = 0$$
$$(5x+4)(5x-4) = 0$$
$$5x + 4 = 0 \qquad 5x - 4 = 0$$
$$x = -\frac{4}{5} \qquad x = \frac{4}{5}$$

77.
$$5x^2 - 6 = 0$$
$$(\sqrt{5}x + \sqrt{6})(\sqrt{5}x - \sqrt{6}) = 0$$
$$\sqrt{5}x + \sqrt{6} = 0 \qquad \sqrt{5}x - \sqrt{6} = 0$$
$$x = -\frac{\sqrt{6}}{\sqrt{5}} \qquad x = \frac{\sqrt{6}}{\sqrt{5}}$$
$$\text{or } x = -\frac{\sqrt{6}}{\sqrt{5}}\cdot\frac{\sqrt{5}}{\sqrt{5}} = -\frac{\sqrt{30}}{5} \qquad \text{or } x = \frac{\sqrt{6}}{\sqrt{5}}\cdot\frac{\sqrt{5}}{\sqrt{5}} = \frac{\sqrt{30}}{5}$$

79. $x^2 + 10x + 25 = 0$
$$(x+5)(x+5) = 0$$
$$x + 5 = 0$$
$$x = -5$$

81.
$$\frac{x^2}{3} - \frac{4x}{3} - 7 = 0$$

multiplying both side y 3: $x^2 - 4x - 21 = 0$

$$(x + 3)(x - 7) = 0$$

$$x = -3, x = 7$$

83. $3x^3 - 14x^2 - 5x = 0$

$$x(3x^2 - 14x - 5) = 0$$

$$x(3x + 1)(x - 5) = 0$$

$$x = 0, x = -\frac{1}{3}, x = 5$$

85. $2x^3 - 5x^2 - 3x = 0$

$$x(2x^2 - 5x - 3) = 0$$

$$x(2x + 1)(x - 3) = 0$$

$$x = 0, x = -\frac{1}{2}, x = 3$$

87. $(2x^2 + 5x - 25)(5x^2 + 9x - 2) = 0$

$$(2x - 5)(x + 5)(5x - 1)(x + 2) = 0$$

$$x = \frac{5}{2}, x = -5, x = \frac{1}{5}, x = -2$$

89. $(4x^2 - 25)(x^2 - 2) = 0$

$$(2x + 5)(2x - 5)(x + \sqrt{2})(x - \sqrt{2}) = 0$$

$$x = \pm\frac{5}{2} \qquad x = \pm\sqrt{2}$$

91 through 99: We use the SIGN method illustrated in Example 2.6.

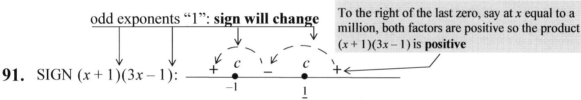

odd exponents "1": **sign will change**

To the right of the last zero, say at x equal to a million, both factors are positive so the product $(x + 1)(3x - 1)$ is **positive**

91. SIGN $(x + 1)(3x - 1)$:

From the above SIGN chart we can read off where $(x + 1)(3x - 1) > 0$: $(-\infty, -1) \cup \left(\frac{1}{3}, \infty\right)$

The "+" intervals

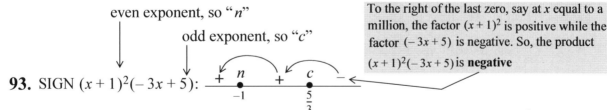

even exponent, so "n"

odd exponent, so "c"

To the right of the last zero, say at x equal to a million, the factor $(x + 1)^2$ is positive while the factor $(-3x + 5)$ is negative. So, the product $(x + 1)^2(-3x + 5)$ is **negative**

93. SIGN $(x + 1)^2(-3x + 5)$:

From the above SIGN chart we can read off where $(x + 1)^2(-3x + 5) \geq 0$: $\left(-\infty, \frac{5}{3}\right]$

Note that we included the points -1 and $\frac{5}{3}$ to accomodate the $\geq$

greater than or **equal** to 0

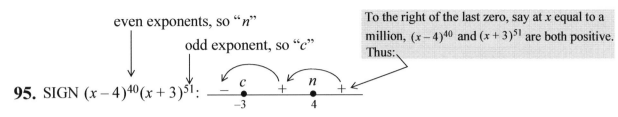

95. SIGN $(x-4)^{40}(x+3)^{51}$:

From the above SIGN chart we can read off where $(x-4)^{40}(x+3)^{51} \leq 0$: $(-\infty, -3] \cup \{4\}$

We included -3 since $(x-4)^{40}(x+3)^{51}$ is equal to 0 at that point

We also had to include 40 for the same reason.

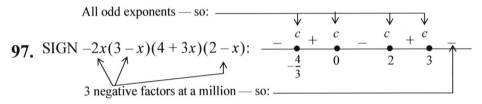

97. SIGN $-2x(3-x)(4+3x)(2-x)$:

3 negative factors at a million — so:

Reading off the "+" intervals
we arrive at the solution of $-2x(3-x)(4+3x)(2-x) > 0$: $(-\frac{4}{3}, 0) \cup (2, 3)$

99. Factor: $\begin{array}{l} x(3x^2 - 14x - 5) \geq 0 \\ x(3x+1)(x-5) \geq 0 \end{array}$ and then SIGN:

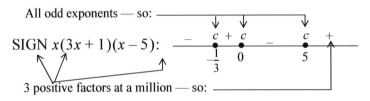

All odd exponents — so:

SIGN $x(3x+1)(x-5)$:

3 positive factors at a million — so:

Reading off the "+" intervals, along
with the zeros, we arrive at the solution of $3x^3 - 14x^2 - 5x \geq 0$: $[-\frac{1}{3}, 0] \cup [5, \infty)$

101.   If: $f(x) = x^2 - x + 2$
Then: $f(-2) = (-2)^2 - (-2) + 2$
$= 4 + 2 + 2 = 8$

103. $f(x) = x^2 - x + 2$

$f\boxed{x+1} = \boxed{x+1}^2 - \boxed{x+1} + 2$

$f(x+1) = (x+1)^2 - (x+1) + 2$
$= x^2 + 2x + 1 - x - 1 + 2$
$= x^2 + x + 2$

105. $f(x) = x^2 - x + 2 \longleftrightarrow f\boxed{} = \boxed{}^2 - \boxed{} + 2$

$$f\boxed{(x+h)} - f(x) = \boxed{(x+h)}^2 - \boxed{(x+h)} + 2 - (x^2 - x + 2)$$

$$= x^2 + 2xh + h^2 - x - h + 2 - x^2 + x - 2 = 2xh + h^2 - h$$

107. If: $f(x) = -2x^2 - 3x + 1$

Then: $f(-2) = -2(-2)^2 - 3(-2) + 1$

$$= -2 \cdot 4 + 6 + 1 = -1$$

109. $f(x) = -2x^2 - 3x + 1$

$$f\,\boxed{x+1} = -2\,\boxed{x+1}^2 - 3\,\boxed{x+1} + 1$$

$$f(x+1) = -2(x+1)^2 - 3(x+1) + 1$$

$$= -2(x^2 + 2x + 1) - 3x - 3 + 1$$

$$= -2x^2 - 4x - 2 - 3x - 2$$

$$= -2x^2 - 7x - 4$$

111. $f(x) = -2x^2 - 3x + 1 \longrightarrow f\boxed{} = -2\boxed{}^2 - 3\boxed{} + 1$

$$f(x+h) - f(x) = -2(x+h)^2 - 3(x+h) + 1 - (-2x^2 - 3x + 1)$$

$$= -2(x^2 + 2xh + h^2) - 3x - 3h + 1 + 2x^2 + 3x - 1$$

$$= -2x^2 - 4xh - 2h^2 - 3h + 2x^2 = -4xh - 2h^2 - 3h$$

113. $f(x) = \begin{cases} x - 3 & \text{if } x < 0 \leftrightarrow \boxed{-5 \quad -1}: f(-5) = -5 - 3 = -8, \ f(-1) = -1 - 3 = -4 \\ 2x & \text{if } x \geq 0 \leftrightarrow \boxed{0 \ 2 \ 10}: f(0) = 0, f(2) = 4, f(10) = 20 \end{cases}$

115. $f(x) = \begin{cases} 2x & \text{if } x < 0 \longleftrightarrow \boxed{-5 \quad -1}: f(-5) = -10, \ f(-1) = -1 \\ -x & \text{if } 0 \leq x < 5 \longleftrightarrow \boxed{0 \quad 2}: f(0) = 0, f(2) = -2 \\ x^2 & \text{if } x \geq 5 \longleftrightarrow \boxed{10}: f(10) = 100 \end{cases}$

§2.2. Limits
Page 137

1. $\lim\limits_{x \to 3} (x^2 - 5) = 3^2 - 5 = 9 - 5 = 4$

3. $\lim\limits_{x \to 3} \dfrac{x^2 - 5}{x + 3} = \dfrac{3^2 - 5}{3 + 3} = \dfrac{4}{6} = \dfrac{2}{3}$

denominator does not approach 0 as x approaches 3 so simply evaluate the expression at 3.

5. $\lim\limits_{x \to 5} \dfrac{x - 5}{x + 5} = \dfrac{5 - 5}{5 + 5} = \dfrac{0}{10} = 0$

denominator does not approach 0 as x approaches 5 so simply evaluate the expression at 5.

7. $\lim\limits_{x \to 5} \dfrac{x^3 - 25x}{x - 5} = \lim\limits_{x \to 5} \dfrac{x(x^2 - 25)}{x - 5} = \lim\limits_{x \to 5} \dfrac{x(x + 5)(x - 5)}{x - 5} = \lim\limits_{x \to 5} x(x + 5) = 5(5 + 5) = 50$

can't evaluate the expression at 5

remember to include the limit sign until you perform the limit operation

9. $\lim\limits_{x \to 5} \dfrac{x^2 - 25}{x^2 - 4x - 5} = \lim\limits_{x \to 5} \dfrac{(x + 5)(x - 5)}{(x + 1)(x - 5)} = \lim\limits_{x \to 5} \dfrac{(x + 5)}{(x + 1)} = \dfrac{10}{6} = \dfrac{5}{3}$

"$\dfrac{0}{0}$" type

include the limit sign till you take the limit

11. $\lim\limits_{x \to -2} \dfrac{x^2 - 4}{x + 2} = \lim\limits_{x \to -2} \dfrac{(x + 2)(x - 2)}{x + 2} = \lim\limits_{x \to -2} (x - 2) = -4$

"$\dfrac{0}{0}$" type

include the limit sign till you take the limit

13. $\lim\limits_{x \to 3} \dfrac{x^2 - 2x - 3}{x^2 - x - 6} = \lim\limits_{x \to 3} \dfrac{(x - 3)(x + 1)}{(x - 3)(x + 2)} = \lim\limits_{x \to 3} \dfrac{x + 1}{x + 2} = \dfrac{4}{5}$

15. $\lim\limits_{x \to -3} \dfrac{x^2 + 4x + 3}{x^2 + 3x} = \lim\limits_{x \to -3} \dfrac{(x + 3)(x + 1)}{x(x + 3)} = \lim\limits_{x \to -3} \dfrac{(x + 1)}{x} = \dfrac{-2}{-3} = \dfrac{2}{3}$

17. $\lim\limits_{x \to 4} \dfrac{x^2 - 8x + 16}{x^2 - 4x} = \lim\limits_{x \to 4} \dfrac{(x - 4)(x - 4)}{x(x - 4)} = \lim\limits_{x \to 4} \dfrac{(x - 4)(x - 4)}{x(x - 4)} = \lim\limits_{x \to 4} \dfrac{(x - 4)}{x} = \dfrac{0}{4} = 0$

19. $\displaystyle\lim_{x\to-1}\frac{2x^3+5x^2+3x}{x^2-3x-4}=\lim_{x\to-1}\frac{x(2x^2+5x+3)}{(x+1)(x-4)}=\lim_{x\to-1}\frac{x(2x+3)(x+1)}{(x+1)(x-4)}$

$$=\lim_{x\to-1}\frac{x(2x+3)}{(x-4)}=\frac{(-1)(1)}{-5}=\frac{1}{5}$$

21. $\displaystyle\lim_{x\to-2}\frac{(x+2)(x-1)}{(x^2+4)(x^2-4)}=\lim_{x\to-2}\frac{(x+2)(x-1)}{(x^2+4)(x+2)(x-2)}=\lim_{x\to-2}\frac{(x-1)}{(x^2+4)(x-2)}=\frac{-3}{-32}=\frac{3}{32}$

23. $\displaystyle\lim_{x\to0}\frac{x^3+2x^2-3x}{x^3-2x^2-15x}=\lim_{x\to0}\frac{x(x^2+2x-3)}{x(x^2-2x-15)}=\lim_{x\to0}\frac{x^2+2x-3}{x^2-2x-15}=\frac{-3}{-15}=\frac{1}{5}$

No need to try to factor the numerator or the denominator: the denominator does not approach 0 as x approaches 0. Simply evaluate the expression at 0.

25. $\displaystyle\lim_{x\to-1}\frac{\frac{1}{x}+1}{x^2-1}=\lim_{x\to-1}\frac{\frac{x+1}{x}}{(x+1)(x-1)}=\lim_{x\to-1}\frac{x+1}{x}\cdot\frac{1}{(x+1)(x-1)}=\lim_{x\to-1}\frac{1}{x(x-1)}$

we have to get rid of the "zero-denominator." So

invert and multiply

$$=\frac{1}{(-1)(-2)}=\frac{1}{2}$$

27. $\displaystyle\lim_{x\to2}\frac{\frac{x^2}{x-1}-4}{\frac{1}{x+2}-\frac{1}{4}}=\lim_{x\to2}\frac{\frac{x^2-4(x-1)}{x-1}}{\frac{4-(x+2)}{4(x+2)}}=\lim_{x\to2}\frac{\frac{x^2-4x+4}{x-1}}{\frac{-x+2}{4(x+2)}}$

we have to get rid of the "zero-denominator." So

invert and multiply

$$=\lim_{x\to2}\frac{\frac{(x-2)(x-2)}{x-1}}{\frac{-(x-2)}{4(x+2)}}=\lim_{x\to2}\frac{(x-2)(x-2)}{x-1}\cdot\frac{4(x+2)}{-(x-2)}$$

$$=\lim_{x\to2}\frac{(x-2)4(x+2)}{-(x-1)}=\frac{0}{-1}=0$$

29.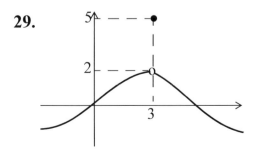

As x approaches 3 **from either side**, the function values approach 2. So: $\displaystyle\lim_{x\to3}f(x)=2$.

Since $f(3)=5\neq\displaystyle\lim_{x\to3}f(x)$, the function is not continuous at 3, it has a removable discontinuity at that point.

31.

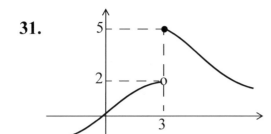

As *x* approach 3 from the left, the function values approach 2. On the other hand, as *x* approach 3 from the right, the function values approach 5.
So, the limit does not exist and the function has a jump discontinuity at 3.

33. $f(x) = \begin{cases} x+2 & \text{if } x<2 \\ x^2 & \text{if } x \geq 2 \end{cases}$ $\longleftrightarrow$ As *x* approaches 2 from the left, $f(x)$ approaches 4.
$\longleftrightarrow$ As *x* approaches 2 from the right, $f(x)$ also approaches 4.

In addition: $f(2) = 4$.

Since $\lim\limits_{x \to 2} f(x) = f(2) = 4$, the function is continuous at 2.

35. $f(x) = \begin{cases} x+2 & \text{if } x<2 \\ 2 & \text{if } x = 2 \\ x^2 & \text{if } x>2 \end{cases}$ $\longleftrightarrow$ As *x* approaches 2 from the left, $f(x)$ approaches 4.

$\longleftrightarrow$ As *x* approaches 2 from the right, $f(x)$ also approaches 4.

Since $\lim\limits_{x \to 2^-} f(x) = \lim\limits_{x \to 2^+} f(x) = 4$, $\lim\limits_{x \to 2} f(x) = 4$. But $f(2) = 2$.

Hence, f is not continuous at 2 (it has a removable discontinuity at that point).

37. Many possible answers. Here is one of them:

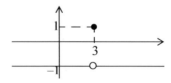

39. Many possible answers. Here is one of them:

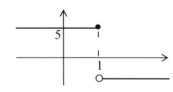

§2.3. Tangent Lines and the Derivative
Page 147

1.

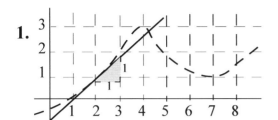
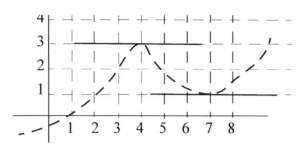

A 1-unit change along the x-axis results in a 1-unit change along the y-axis. So the represented tangent line appears to have a slope of $\frac{\Delta y}{\Delta x} = \frac{1}{1} = 1$. Since the derivative represents the slope of the tangent line: $f'(2) = 1$.

The slope of the tangent lines at $x = 4$ and at $x = 7$ appear to be horizontlal. As such, they have a slope of 0. Consequently:
$$f'(4) = 0 \text{ and } f'(7) = 0$$

3. Invoking the definition: $f'(c) = \lim\limits_{h \to 0} \frac{f(c+h)-f(c)}{h}$ with $f(x) = 5x+1$ and $c = 2$ we have:

$$f(x) = 5x+1 \Rightarrow f(2+h) = \boxed{5(2+h)+1}$$

$$f'(2) = \lim_{h \to 0} \frac{f(2+h)-f(2)}{h} = \lim_{h \to 0} \frac{5(2+h)+1-(5\cdot2+1)}{h}$$

$$= \lim_{h \to 0} \frac{10+5h+1-11}{h} = \lim_{h \to 0} \frac{5h}{h} = \lim_{h \to 0} 5 = 5$$

> Thinking geometrically, it should not be surprising to find that $f'(2) = 5$:
> The tangent line to the graph of the line $f(x) = 5x+1$ is the line itself: a line of slope 5.

5.
$$f(x) = 4x^2 \Rightarrow f(2+h) = \boxed{4(2+h)^2} \qquad f(2) = 4\cdot2^2 = 16$$

$$f'(2) = \lim_{h \to 0} \frac{f(2+h)-f(2)}{h} = \lim_{h \to 0} \frac{4(2+h)^2-16}{h}$$

$$= \lim_{h \to 0} \frac{4(4+4h+h^2)-16}{h} = \lim_{h \to 0} \frac{16+16h+4h^2-16}{h}$$

$$= \lim_{h \to 0} \frac{16h+4h^2}{h}$$

$$= \lim_{h \to 0} \frac{h(16+4h)}{h}$$

$$= \lim_{h \to 0} (16+4h) = 16$$

7. $f(x) = -x^2 + 3x - 1 \Rightarrow f(2+h) = -(2+h)^2 + 3(2+h) - 1$

$f(2) = -2^2 + 3 \cdot 2 - 1$
$= -4 + 6 - 1$
$= -1$

$f'(2) = \lim_{h \to 0} \dfrac{f(2+h) - f(2)}{h} = \lim_{h \to 0} \dfrac{-(2+h)^2 + 3(2+h) - 1 - 1}{h}$

$= \lim_{h \to 0} \dfrac{-(4 + 4h + h^2) + 6 + 3h - 2}{h}$

$= \lim_{h \to 0} \dfrac{-4 - 4h - h^2 + 4 + 3h}{h}$

$= \lim_{h \to 0} \dfrac{-4h - h^2 + 3h}{h} = \lim_{h \to 0} \dfrac{h(-4 - h + 3)}{h}$

$= \lim_{h \to 0} (-4 - h + 3) = -1$

9. Since $f(x) = 55$ is a constant function, its value at **any** x is 55. In particular, $f(2) = 55$ and $f(2+h) = 55$ for every h. Bringing us to:

$f'(2) = \lim_{h \to 0} \dfrac{f(2+h) - f(2)}{h} = \lim_{h \to 0} \dfrac{55 - 55}{h} = \lim_{h \to 0} \dfrac{0}{h} = 0$

for every $h \neq 0$: $\dfrac{0}{h} = 0$

Thinking geometrically, it should not be surprising to find that $f'(2) = 0$:
The tangent line to the graph of the horizontal line $f(x) = 55$ is the line itself: a line of slope 0.

11. $f(x) = x^3 + x + 1 \Rightarrow f(2+h) = (2+x)^3 + (2+h) + 1 = (8 + 12h + 6h^2 + h^3) + h + 3$

$(2+h)^3 = (2+h)^2(2+h)$
$= (4 + 4h + h^2)(2+h)$

$\begin{array}{r} 4 + 4h + h^2 \\ 2 + h \\ \hline 8 + 8h + 2h^2 \\ 4h + 4h^2 + h^3 \\ \hline 8 + 12h + 6h^2 + h^3 \end{array}$

$f'(2) = \lim_{h \to 0} \dfrac{f(2+h) - f(2)}{h} = \lim_{h \to 0} \dfrac{(8 + 12h + 6h^2 + h^3 + h + 3) - (2^3 + 2 + 1)}{h}$

$= \lim_{h \to 0} \dfrac{h^3 + 6h^2 + 12h + 11 - 11}{h} = \lim_{h \to 0} \dfrac{h^3 + 6h^2 + 13h}{h}$

$= \lim_{h \to 0} \dfrac{h(h^2 + 6h + 13)}{h}$

$= \lim_{h \to 0} (h^2 + 6h + 13) = 13$

13. For $f(x) = -5x - 4000$:

$$f'(x) = \lim_{h \to 0} \frac{f(x+h) - f(x)}{h} = \lim_{h \to 0} \frac{-5(x+h) - 4000 - (-5x - 4000)}{h}$$

$$= \lim_{h \to 0} \frac{-5x - 5h - 4000 + 5x + 4000}{h} = \lim_{h \to 0} \frac{-5h}{h} = -5$$

15. For $f(x) = 3x^2$:

$$f'(x) = \lim_{h \to 0} \frac{f(x+h) - f(x)}{h} = \lim_{h \to 0} \frac{3(x+h)^2 - 3x^2}{h}$$

$$= \lim_{h \to 0} \frac{3(x^2 + 2xh + h^2) - 3x^2}{h} = \lim_{h \to 0} \frac{3x^2 + 6xh + 3h^2 - 3x^2}{h}$$

$$= \lim_{h \to 0} \frac{6xh + 3h^2}{h}$$

$$= \lim_{h \to 0} \frac{h(6x + 3h)}{h}$$

$$= \lim_{h \to 0} (6x + 3h) = 6x$$

17. For $f(x) = -2x^2 + x - 2$:

$$f'(x) = \lim_{h \to 0} \frac{f(x+h) - f(x)}{h} = \lim_{h \to 0} \frac{-2(x+h)^2 + (x+h) - 2 - (-2x^2 + x - 2)}{h}$$

$$= \lim_{h \to 0} \frac{-2(x^2 + 2xh + h^2) + x + h - 2 + 2x^2 - x + 2}{h}$$

$$= \lim_{h \to 0} \frac{-2x^2 - 4xh - 2h^2 + h + 2x^2}{h}$$

$$= \lim_{h \to 0} \frac{-4xh - 2h^2 + h}{h} = \lim_{h \to 0} \frac{h(-4x - 2h + 1)}{h}$$

$$= \lim_{h \to 0} (-4x - 2h + 1) = -4x + 1$$

19. For $f(x) = 101$: $f'(x) = \lim_{h \to 0} \frac{f(x+h) - f(x)}{h} = \lim_{h \to 0} \frac{101 - 101}{h} = \lim_{h \to 0} \frac{0}{h} = 0$

21. The slope m of the tangent line to the graph of the function $f(x) = -3x$ at $x = 5$ is:

$$f'(5) = \lim_{h \to 0} \frac{f(5+h) - f(5)}{h} = \lim_{h \to 0} \frac{-3(5+h) - 15}{h} = \lim_{h \to 0} \frac{-3h}{h} = \lim_{h \to 0} -3 = -3.$$

So, the tangent line is of the form $y = -3x + b$. Since the tangent line touches the graph of $f(x) = -3x$ at $x = 5$, the point $(5, f(5)) = (5, -15)$ lies on the tangent line; bringing us to:

$$(5, -15) \to y = -3x + b$$

$$-15 = 3(5) + b \Rightarrow b = -30$$

Conclusion: $y = -3x - 30$ is the tangent line to the graph of $f(x) = -3x$ at $x = 5$.

23. The slope m of the tangent line to the graph of the function $f(x) = x^2 + 2x$ at $x = 0$ is:

$$f'(0) = \lim_{h \to 0} \frac{f(0+h) - f(0)}{h} = \lim_{h \to 0} \frac{f(h) - 0}{h} = \lim_{h \to 0} \frac{h^2 + 2h}{h} = \lim_{h \to 0} (h+2) = 2$$

So, the tangent line is of the form $y = 2x + b$. Since the tangent line touches the graph of $f(x) = x^2 + 2x$ at $x = 0$, the point $(0, f(0)) = (0, 0)$ lies on the tangent line; bringing us to:

$$(0,0) \to y = 2x + b$$
$$0 = 2 \cdot 0 + b \Rightarrow b = 0$$

Conclusion: $y = 2x$ is the tangent line to the graph of $f(x) = x^2 + 2x$ at $x = 0$.

25. The slope m of the tangent line to the graph of the function $f(x) = x^2 + 2x$ at $x = 2$ is:

$$f'(2) = \lim_{h \to 0} \frac{f(2+h) - f(2)}{h} = \lim_{h \to 0} \frac{(2+h)^2 + 2(2+h) - 8}{h} = \lim_{h \to 0} \frac{4 + 4h + h^2 + 4 + 2h - 8}{h}$$
$$= \lim_{h \to 0} \frac{6h + h^2}{h}$$
$$= \lim_{h \to 0} (6+h) = 6$$

So, the tangent line is of the form $y = 6x + b$. Since the tangent line touches the graph of $f(x) = x^2 + 2x$ at $x = 2$, the point $(2, f(2)) = (2, 8)$ lies on the tangent line; bringing us to:

$$(2,8) \to y = 6x + b$$
$$8 = 6 \cdot 2 + b \Rightarrow b = -4$$

Conclusion: $y = 6x - 4$ is the tangent line to the graph of $f(x) = x^2 + 2x$ at $x = 2$.

27. The slope m of the tangent line to the graph of the constant function $f(x) = 11$ at $x = -9$ is:

$$f'(0) = \lim_{h \to 0} \frac{f(-9+h) - f(-9)}{h} = \lim_{h \to 0} \frac{11 - 11}{h} = \lim_{h \to 0} \frac{0}{h} = 0. \text{ So, the tangent line is of the}$$
form $y = 0 \cdot x + b$.

Since the tangent line touches the graph of $f(x) = 11$ at $x = -9$, the point $(-9, f(-9)) = (-9, 11)$ lies on the tangent line; bringing us to:

$$(-9, 11) \to y = 0 \cdot x + b$$
$$11 = 0(-9) + b \Rightarrow b = 11$$

Conclusion: $y = 11$ is the tangent line to the graph of $f(x) = 11$ at $x = -9$.

29. (A geometrical argument.) The graph of $f(x) = x$ is a line of slope 1. The tangent line to any point on that line is the line itself. It follows that $f'(x) = 1$ for every x.

(An analytical argument.) For $f(x) = x$:

$$f'(x) = \lim_{h \to 0} \frac{f(x+h) - f(x)}{h} = \lim_{h \to 0} \frac{x+h-x}{h} = \lim_{h \to 0} \frac{h}{h} = \lim_{h \to 0} 1 = 1$$

31. (A geometrical argument.) The graph of $f(x) = mx + b$ is a line of slope m. The tangent line to any point on that line is the line itself. It follows that $f'(x) = m$ for every x.

(An analytical argument.) For $f(x) = x$:

$$f'(x) = \lim_{h \to 0} \frac{f(x+h) - f(x)}{h} = \lim_{h \to 0} \frac{m(x+h) + b - (mx + b)}{h}$$

$$= \lim_{h \to 0} \frac{mx + mh + b - mx - b}{h} = \lim_{h \to 0} \frac{mh}{h} = \lim_{h \to 0} m = m$$

33.

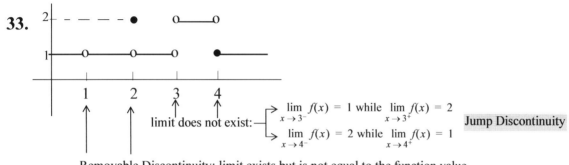

$$\text{limit does not exist:} \begin{cases} \rightarrow \lim_{x \to 3^-} f(x) = 1 \text{ while } \lim_{x \to 3^+} f(x) = 2 \\ \rightarrow \lim_{x \to 4^-} f(x) = 2 \text{ while } \lim_{x \to 4^+} f(x) = 1 \end{cases}$$ Jump Discontinuity

Removable Discontinuity: limit exists but is not equal to the function value

Not differentiable at 1, 2, 3, and 4 (Theorem 2.5)

35. For $f(x) = \begin{cases} 2x + 2 & \text{if } x \le 2 \\ 3x & \text{if } x > 2 \end{cases}$:

$$\lim_{x \to 2^-} \frac{f(2+h) - f(2)}{h} = \lim_{x \to 2^-} \frac{2(2+h) + 2 - 6}{h} = \lim_{x \to 2^-} \frac{4 + 2h - 4}{h} = \lim_{x \to 2^-} \frac{2h}{h} = 2$$

while: $\lim_{x \to 2^+} \frac{f(2+h) - f(2)}{h} = \lim_{x \to 2^+} \frac{3(2+h) - 6}{h} = \lim_{x \to 2^+} \frac{6 + 3h - 6}{h} = \lim_{x \to 2^+} \frac{3h}{h} = 3$

It follows that $f'(x) = \lim_{x \to 2} \frac{f(2+h) - f(2)}{h}$ does not exist (left-hand limit does not equal right-hand limit).

37.

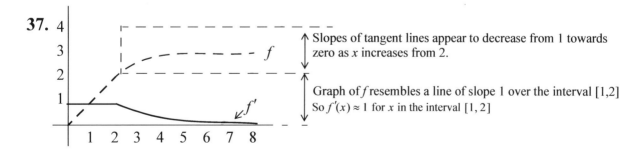

Slopes of tangent lines appear to decrease from 1 towards zero as x increases from 2.

Graph of f resembles a line of slope 1 over the interval $[1,2]$
So $f'(x) \approx 1$ for x in the interval $[1, 2]$

39. Let $h = x - c$. As $h \to 0: x \to c$ and $c + h \to x$. Thus:

$$f'(c) = \lim_{h \to 0} \frac{f(c+h) - f(c)}{h} = \lim_{x \to c} \frac{f(x) - f(c)}{x - c}$$

§2.4. Differentiation Formulas
Page 157

1. Since the derivative of x is 1 and that the derivative of a constant is 0: $f'(x) = (2x + 7)' = 2$

3. $f'(x) = (3x^5 + 4x^3 - 7)' = 15x^4 + 12x$ **5.** $g'(x) = (2x^{-3} + 4)' = -6x^{-4} = -\dfrac{6}{x^4}$

7. $g'(x) = (7x^3 + 5x^2 - 4x + x^{-4} + 1)' = 21x^2 + 10x - 4 - 4x^{-5} = 21x^2 + 10x - 4 - \dfrac{4}{x^5}$

9.

$$\overset{\text{get to powers of } x}{\downarrow} \qquad \overset{\text{then differentiate}}{\downarrow}$$

$$g'(x) = \left(x + \frac{2}{x} + \frac{3}{x^2}\right)' = (x + 2x^{-1} + 3x^{-2})' = 1 - 2x^{-2} - 6x^{-3} = 1 - \frac{2}{x^2} - \frac{6}{x^3}$$

11. $k'(x) = \left(\dfrac{x^5 + 3x - 5}{x^3}\right)' = \left(\dfrac{x^5}{x^3} + \dfrac{3x}{x^3} - \dfrac{5}{x^3}\right)' = (x^2 + 3x^{-2} - 5x^{-3})'$

$$= 2x - 6x^{-3} + 15x^{-4} = 2x - \frac{6}{x^3} + \frac{15}{x^4}$$

13.

$$K'(x) = [(x^3 + 2x)(3x^3 + 2x + 3)]'$$
$$= (3x^6 + 8x^4 + 3x^3 + 4x^2 + 6x)' = 18x^5 + 32x^3 + 9x^2 + 8x + 6$$

$$
\begin{array}{r}
3x^3 + 2x + 3 \\
x^3 + 2x \\
\hline
3x^6 + 2x^4 + 3x^3 \\
6x^4 + 4x^2 + 6x \\
\hline
3x^6 + 8x^4 + 3x^3 + 4x^2 + 6x
\end{array}
$$

15. $F'(x) = \left(\dfrac{3x^2 + 2x - 5}{x}\right)' = \left(\dfrac{3x^2}{x} + \dfrac{2x}{x} - \dfrac{5}{x}\right)' = (3x + 2 - 5x^{-1})' = 3 + 5x^{-2} = 3 + \dfrac{5}{x^2}$

17. $F'(x) = \left(\dfrac{3x^2 + 2x - 5}{x + 4}\right)' = \dfrac{(x+4)(3x^2 + 2x - 5)' - (3x^2 + 2x - 5)(x + 4)'}{(x + 4)^2}$

need to use the quotient rule:

$$\left[\frac{f(x)}{g(x)}\right]' = \frac{g(x) \cdot f'(x) - f(x) \cdot g'(x)}{[g(x)]^2}$$

$$= \frac{(x + 4)(6x + 2) - (3x^2 + 2x - 5)(1)}{(x + 4)^2}$$

$$= \frac{6x^2 + 24x + 2x + 8 - 3x^2 - 2x + 5}{(x + 4)^2} = \frac{3x^2 + 24x + 13}{(x + 4)^2}$$

19.
$$F'(x) = \left(\frac{5}{3x^2+1}\right)' = \frac{(3x^2+1)(5)' - 5(3x^2+1)'}{(3x^2+1)^2}$$

using the quotient rule:

$$\left[\frac{f(x)}{g(x)}\right]' = \frac{g(x)\cdot f'(x) - f(x)\cdot g'(x)}{[g(x)]^2}$$

$$= \frac{(3x^2+1)\cdot 0 - 5(6x)}{(3x^2+1)^2} = -\frac{30x}{(3x^2+1)^2}$$

21. $H'(x) = \left(\dfrac{x}{2x+1} + \dfrac{x}{3x-1}\right)' = \left(\dfrac{x}{2x+1}\right)' + \left(\dfrac{x}{3x-1}\right)'$

$$= \frac{(2x+1)x' - x(2x+1)'}{(2x+1)^2} + \frac{(3x-1)x' - x(3x-1)'}{(3x-1)^2}$$

$$= \frac{(2x+1)(1) - x(2)}{(2x+1)^2} + \frac{(3x-1)(1) - x(3)}{(3x-1)^2}$$

$$= \frac{2x+1-2x}{(2x+1)^2} + \frac{3x-1-3x}{(3x-1)^2} = \frac{1}{(2x+1)^2} - \frac{1}{(3x-1)^2}$$

$$= \frac{(3x-1)^2 - (2x+1)^2}{(2x+1)^2(3x^2-x)^2}$$

$$= \frac{9x^2 - 6x + 1 - (4x^2 + 4x + 1)}{(2x+1)^2(3x^2-x)^2}$$

$$= \frac{5x^2 - 10x}{(2x+1)^2(3x^2-x)^2}$$

23. $\left[\left(\dfrac{x}{3x+1}\right)(x^2+2x)\right]' = \left(\dfrac{x^3+2x^2}{3x+1}\right)' = \dfrac{(3x+1)(x^3+2x^2)' - (x^3+2x^2)(3x+1)'}{(3x+1)^2}$

$$= \frac{(3x+1)(3x^2+4x) - (x^3+2x^2)(3)}{(3x+1)^2}$$

$$= \frac{9x^3 + 15x^2 + 4x - 3x^3 - 6x^2}{(3x+1)^2}$$

$$= \frac{6x^3 + 9x^2 + 4x}{(3x+1)^2}$$

Note: You can start with the product rule:

$$\left[\left(\frac{x}{3x+1}\right)(x^2+2x)\right]' = \left(\frac{x}{3x+1}\right)(x^2+2x)' + (x^2+2x)\left(\frac{x}{3x+1}\right)'$$

but that approach does not turn out to be any easier.

25. $f''(x) = (x^3 - 6x^2 + 12x)'' = (3x^2 - 12x + 12)' = 6x - 12$

differentiate once   then differentiate again

27. $f''(x) = \left(x^3 + 2x^2 - \dfrac{1}{x}\right)'' = (x^3 + 2x^2 - x^{-1})'' = (3x^2 + 4x + x^{-2})' = 6x + 4 - 2x^{-3}$

differentiate once

differentiate again

$= 6x + 4 - \dfrac{2}{x^3}$

29. We are given that $f'(1) = 6$ and that $g'(1) = 2$. So:

$$(f+g)'(1) = f'(1) + g'(1) = 6 + 2 = 8$$

31. We are given that $f'(1) = 6$, $g'(1) = 2$, $f(1) = 3$, and $g(1) = 2$. So:

$$\left(\dfrac{f}{g}\right)'(1) = \dfrac{g(1)f'(1) - f(1)g'(1)}{[g(1)]^2} = \dfrac{(2)(6) - (3)(2)}{2^2} = \dfrac{6}{4} = \dfrac{3}{2}$$

33. We are given that $f'(2) = 0$, $g'(2) = 2$, and $h'(2) = 1$ So:

$$(f + g + h)'(2) = f'(2) + g'(2) + h'(2) = 0 + 2 + 1 = 3$$

35. We are given that $f'(2) = 0$, $g'(2) = 2$, $h'(2) = 1$, $f(2) = 6$, and $g(2) = 5$ So:

$$(fg + h)'(2) = (fg)'(2) + h'(2) = f(2)g'(2) + g(2)f'(2) + h'(2)$$
$$= (6)(2) + (5)(0) + 1 = 13$$

37. We are given that $f'(1) = 6$, $g'(1) = 2$, $h'(1) = 1$, $f(1) = 3$, $g(1) = 2$, and $h(1) = 6$

So: $\left(\dfrac{fg}{h}\right)'(1) = \dfrac{h(1)(fg)'(1) - (fg)(1)h'(1)}{[h(1)]^2} = \dfrac{6[f(1)g'(1) + g(1)f'(1)] - f(1)g(1) \cdot 1}{6^2}$

$$= \dfrac{6[(3)(2) + (2)(6)] - (3)(2)}{36}$$

$$= \dfrac{(6)(18) - 6}{36} = \dfrac{17}{6}$$

39. We are given that $g(1) = 2$, $g'(2) = 2$, $h'(2) = 1$, $g(2) = 5$, and $h(2) = 2$. So:

$$\left(\dfrac{g}{h} + g\right)'(t) = \left(\dfrac{g}{h} + g\right)'(2) = \left(\dfrac{g}{h}\right)'(2) + g'(2) = \dfrac{h(2)g'(2) - g(2)h'(2)}{[h(2)]^2} + 2$$

since $t = g(1)$

$$= \dfrac{(2)(2) - (5)(1)}{2^2} + 2 = -\dfrac{1}{4} + 2 = \dfrac{7}{4}$$

41. We are give that $f(0) = 1$ and that $g(1) = 2$. So: $g[f(0)] = g(1) = 2$. We are also given that $f'(2) = 0$, $g'(2) = 2$, $h'(2) = 1$, $f(2) = 6$, $g(2) = 5$, and $h(2) = 2$. So:

$$\left(\frac{f+g+h}{f-g-h}\right)'(x) = \left(\frac{f+g+h}{f-g-h}\right)'(2)$$

$$= \frac{[(f-g-h)(2)][(f+g+h)'(2)] - [(f+g+h)(2)][(f-g-h)'(2)]}{[(f-g-h)(2)]^2}$$

$$= \frac{[f(2)-g(2)-h(2)][f'(2)+g'(2)+h'(2)] - [f(2)+g(2)+h(2)][f'(2)-g'(2)-h'(2)]}{[f(2)-g(2)-h(2)]^2}$$

$$= \frac{(6-5-2)(0+2+1) - (6+5+2)(0-2-2)}{(6-5-2)^2} = \frac{(-1)(3) - (13)(-4)}{(-1)^2} = 49$$

43. We first find the slope m of the tangent line to the graph of $f(x) = 3x^2 - x - 1$ at $x = 1$:

$$f'(x) = (3x^2 - x - 1)' = 6x - 1$$

$$\text{So: } m = f'(1) = 6 \cdot 1 - 1 = 5$$

At this point we know that the tangent line is of the form $y = 5x + b$. Using the fact that the tangent line touches the curve at $(1, f(1)) = (1, 1)$ enables us to determine the value of b:

$$3 \cdot 1^2 - 1 - 1 = 1$$

$$(1, 1) \to y = 5x + b$$

$$1 = 5 \cdot 1 + b \Rightarrow b = -4$$

Conclusion: $y = 5x - 4$ is the tangent line to the graph of $f(x) = x^2 + 2x$ at $x = 1$.

45. We first find the slope m of the tangent line to the graph of $f(x) = \dfrac{x^5 + 2x}{x^4}$ at $x = -1$:

$$f'(x) = \left(\frac{x^5 + 2x}{x^4}\right)' = (x + 2x^{-3})' = 1 - 6x^{-4} = 1 - \frac{6}{x^4}$$

$$\text{So: } m = f'(-1) = 1 - \frac{6}{(-1)^4} = 1 - 6 = -5$$

At this point we know that the tangent line is of the form $y = -5x + b$. Using the fact that the tangent line touches the curve at $(-1, f(-1)) = (-1, -3)$ enables us to determine the value of b:

$$(-1, -3) \to y = -5x + b$$

$$-3 = -5(-1) + b \Rightarrow b = -8$$

Conclusion: $y = -5x - 8$ is the tangent line to the graph of $f(x) = \dfrac{x^5 + 2x}{x^4}$ at $x = -1$.

47. We first find the slope m of the tangent line to the graph of $f(x) = \dfrac{x^2 + 2x}{x - 1}$ at $x = -1$:

$$f'(x) = \left(\frac{x^2 + 2x}{x - 1}\right)' = \frac{(x - 1)(x^2 + 2x)' - (x^2 + 2x)(x - 1)'}{(x - 1)^2} = \frac{(x - 1)(2x + 2) - (x^2 + 2x)}{(x - 1)^2}$$

So: $m = f'(-1) = \dfrac{(-1 - 1)[2(-1) + 2] - [(-1)^2 + 2(-1)]}{(-1 - 1)^2} = \dfrac{(-2)(0) - (1 - 2)}{4} = \dfrac{1}{4}$

At this point we know that the tangent line is of the form $y = \dfrac{1}{4}x + b$. Using the fact that the tangent line touches the curve at $(-1, f(-1)) = (-1, -\dfrac{1}{2})$ enables us to determine b:

$$\left(-1, \tfrac{1}{2}\right): \; y = \frac{1}{4}x + b$$

$$\frac{1}{2} = \frac{1}{4}(-1) + b \Rightarrow b = \frac{1}{2} + \frac{1}{4} = \frac{3}{4}$$

Conclusion: $y = \dfrac{1}{4}x + \dfrac{3}{4}$ is the tangent line to the graph of $f(x) = \dfrac{x^2 + 2x}{x - 1}$ at $x = -1$.

49. To have a horizontal tangent line is to have a **zero derivative**. So, to find where the horizontal tangent lines to the graph of $f(x) = x^3 - 6x^2 + 12x$, we solve the equation:

$$(x^3 - 6x^2 + 12x)' = 0$$
$$3x^2 - 12x + 12 = 0$$
Divide both sides by 3: $x^2 - 4x + 4 = 0$
$$(x - 2)(x - 2) = 0$$
$$x = 2$$

The graph of $f(x) = x^3 - 6x^2 + 12x$ has a horizontal tangent line at $x = 2$.

51. To have a horizontal tangent line is to have a **zero derivative**. So, to find where the horizontal tangent lines to the graph of $f(x) = \dfrac{2}{3}x^3 - \dfrac{1}{2}x^2 - x + 2$, we solve the equation:

$$\left(\frac{2}{3}x^3 - \frac{1}{2}x^2 - x + 2\right)' = 0$$

$$2x^2 - x - 1 = 0$$
$$(2x + 1)(x - 1) = 0$$
$$x = -\frac{1}{2} \text{ and } x = 1$$

The graph of $f(x) = \dfrac{2}{3}x^3 - \dfrac{1}{2}x^2 - x + 2$ has a horizontal tangent line at $x = -\dfrac{1}{2}$ and at $x = 1$.

53.
$$(f+g)'(x) = \lim_{h \to 0} \frac{(f+g)(x+h)-(f+g)(x)}{h}$$

$$= \lim_{h \to 0} \frac{f(x+h)+g(x+h)-f(x)-g(x)}{h}$$

regroup: $$= \lim_{h \to 0} \left[\frac{f(x+h)-f(x)}{h} + \frac{g(x+h)-g(x)}{h} \right]$$

Theorem 2.4(1), page 135: $$= \lim_{h \to 0} \frac{f(x+h)-f(x)}{h} + \lim_{h \to 0} \left[\frac{g(x+h)-g(x)}{h} \right] = f'(x)+g'(x)$$

A similar argument can be used to show that $(f-g)'(x) = f'(x)-g'(x)$.

55. $(x^2)' = \lim_{h \to 0} \frac{(x+h)^2-x^2}{h} = \lim_{h \to 0} \frac{x^2+2xh+h^2-x^2}{h}$

$$= \lim_{h \to 0} \frac{2xh+h^2}{h} = \lim_{h \to 0} \frac{h(2x+h)}{h} = \lim_{h \to 0} (2x+h) = 2x$$

Then: $(x^4)' = (x^2x^2)' = x^2(x^2)' + (x^2)(x^2)' = x^2(2x)+x^2(2x) = 4x^3$

and: $(x^5)' = (x^4x)' = x^4(x)' + (x)(x^4)' = x^4(1)+x(4x^3) = 5x^4$.

---

§2.5. Other Interpretations of the Derivative
Page 167

World Population (billions)

1.

Year	1992	1993	1994	1995	1996	1997	1998	1999	2000
Population	5.45	5.53	5.61	5.69	5.77	5.85	5.92	6.00	6.08

(a) From 1992 to 1995: $\dfrac{5.69-5.45}{1995-1992} = \dfrac{0.24}{3} = 0.08000$ billion per year

(b) From 1995 to 2000: $\dfrac{6.08-5.69}{2000-1995} = \dfrac{0.24}{5} = 0.07800$ billion per year

(c) From 1992 to 2000: $\dfrac{6.08-5.45}{2000-1992} = \dfrac{0.24}{8} = 0.07875$ billion per year

3. For $f(x) = x^3+2x-3$, $a = 0, b = \dfrac{1}{2}$:

(a) and (b): Slope of the line and average rate of change are one and the same; namely:

$$\frac{\Delta y}{\Delta x} = \frac{f\left(\frac{1}{2}\right)-f(0)}{\frac{1}{2}-0} = \frac{\left[\left(\frac{1}{2}\right)^3 + 2 \cdot \frac{1}{2} - 3\right]-(-3)}{\frac{1}{2}} = \frac{9/8}{1/2} = \frac{9}{4}$$

(c) and (d): The slope of the tangent line and the instantaneous rate of change are one and the same; namely: $f'(0) \underset{\uparrow}{=} 2.$

since $f'(x) = 3x^2 + 2$

5. For $f(x) = \dfrac{x^2 + 1}{x}$, $a = 1, b = 3$:

(a) and (b): Slope of the line and average rate of change are one and the same; namely:

$$\frac{\Delta y}{\Delta x} = \frac{f(3) - f(1)}{3 - 1} = \frac{\frac{10}{3} - 2}{2} = \frac{\frac{4}{3}}{2} = \frac{4}{3} \cdot \frac{1}{2} = \frac{2}{3}$$

(c) and (d): The slope of the tangent line and the instantaneous rate of change are one and the same; namely: $f'(1) \underset{\uparrow}{=} 0$

since $f'(x) = \left(\dfrac{x^2 + 1}{x}\right)' = (x + x^{-1})' = 1 - x^{-2} = 1 - \dfrac{1}{x^2}$

7. From the formula $A = \pi r^2$ we have $\dfrac{dA}{dr} = 2\pi r$. To find the rate of change of the area with respect to the radius when $r = 2$ we evaluate the above derivative at 2: $\dfrac{dA}{dr}\bigg|_2 = 2\pi(2) = 4\pi$.

9. (a) We are given the position function $s(t) = -16t^2 + 128t$. Since velocity, $v(t)$, is the rate of change of position with respect to time, we have:

$$v(t) = \frac{ds}{dt} = \frac{d}{dt}(-16t^2 + 128t) = -32t + 128 \text{ feet per second}$$

In particular: $v(2) = -32 \cdot 2 + 128 = 64\dfrac{\text{ft}}{\text{sec}}$.

(b) At the instant at which the ball is at its highest level, it is standing still. Consequently, we set velocity to 0, and solve for the time it takes for the ball to reach its zenith:
$$v(t) = -32t + 128 = 0$$
$$-32t = -128$$
$$t = 4 \text{ seconds}$$
Knowing it reaches its highest level in 4 seconds, we find the maximum height of the ball by evaluating the position function at that time:
$$s(4) = -16 \cdot 4^2 + 128 \cdot 4 = 256 \text{ feet}$$

(c) Setting the position function to zero (ground is reference point), we find how long it takes for the ball to hit the ground:
$$s(t) = -16t^2 + 128t = 0$$
$$-16t(t - 8) = 0$$

ball is at ground level (leaves from ground level) $\rightarrow t = 0$ or $t = 8 \leftarrow$ ball is at ground level (hits the ground)

Knowing that it takes 8 seconds for the ball to hit the ground, the impact velocity is determined by evaluating the velocity function at that time:

$$v(8) = -32 \cdot 8 + 128 = -128 \, \frac{\text{ft}}{\text{sec}}$$

$$\text{Impace speed} = |v(8)| = 128 \frac{\text{ft}}{\text{sec}}$$

11. We are given the monthly cost function: $C(x) = 1000 + 23x - \dfrac{x^2}{25}$.

(a) The company has a monthly fixed cost of \$1000.

(b) $\overline{MC}(x) = \left(1000 + 23x - \dfrac{x^2}{25}\right)' = 23 - \dfrac{2x}{25}$.

(c) $\overline{MC}(100) = 23 - \dfrac{2(100)}{25} = 23 - 8 = \$15$.

(d) The approximate cost to produce the 101th unit.

(e) $C(101) - C(100) = \left[1000 + 23(101) - \dfrac{101^2}{25}\right] - \left[1000 + 23(100) - \dfrac{100^2}{25}\right]$

$$= 23 + \frac{100^2 - 101^2}{25} = \$14.96$$

13. We are given the weekly cost function: $C(x) = 37{,}000 + 22.5x - 0.01x^2$.

(a) The company has a weekly fixed cost of \$37,000.

(b) $\overline{MC}(x) = (37{,}000 + 22.5x - 0.01x^2)' = 22.5 - 0.02x$.

(c) $\overline{MC}(50) = 22.5 - 0.02(50) = \$21.50$.

(d) The approximate cost to produce the 51th unit.

(e) $C(51) - C(50) = [37{,}000 + 22.5(51) - 0.01(51)^2] - [37{,}000 + 22.5(50) - 0.01(50)^2]$

$$= 22.5 + 0.01(50^2 - 51^2) = \$21.49$$

15. (a) $\overline{MR}(x) = (50x + 0.02x^2)' = 50 + 0.04x$

(b) $\overline{MR}(100) = 50 + 0.04(100) = \$54.00$

(c) The approximate revenue from the sale of the 101th unit.

(d) $R(101) - R(100) = [50(101) + 0.02(101)^2] - [50(100) + 0.02(100)^2]$

$$= 50 + 0.02(101^2 - 100^2) = 54.02$$

17. (a) $P(x) = R(x) - C(x) = (65x + 0.12x^2) - (2500 + 50x) = 0.12x^2 + 15x - 2500$

and $\overline{MP}(x) = (0.12x^2 + 15x - 2500)' = 0.24x + 15$.

(b) $\overline{MP}(100) = 0.24(100) + 15 = \$39.00$

(c) The approximate profit from the sale of the 101th unit.

(d) $P(101) - P(100) = [0.12(101)^2 + 15(101) - 2500] - [0.12(100)^2 + 15(100) - 2500]$

$= 0.12(101^2 - 100^2) + 15 = \$39.12$

19.

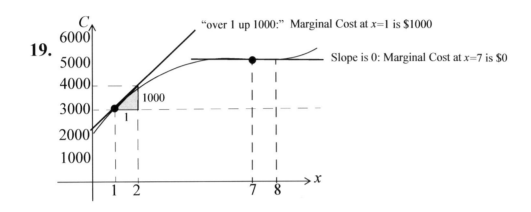

§2.6. Graphing Polynomial Functions
Page 180

1. To find the critical points of $f(x) = \dfrac{3x^4}{2} - 4x^3 - 9x^2$, we determine where $f'(x) = 0$:

$\left(\dfrac{3x^4}{2} - 4x^3 - 9x^2\right)' = 0$

$6x^3 - 12x^2 - 18x = 0$

$6x(x^2 - 2x - 3) = 0$

$6x(x + 1)(x - 3) = 0$

$x = 0, x = -1, x = 3$
Critical Points

Values

For $x = 0$: $f(0) = \dfrac{3(0)^4}{2} - 4(0)^3 - 9(0)^2 = 0$

For $x = -1$: $f(-1) = \dfrac{3(-1)^4}{2} - 4(-1)^3 - 9(-1)^2 \stackrel{?}{=} -\dfrac{7}{2}$

details committed

For $x = 3$: $f(3) = \dfrac{3(3)^4}{2} - 4(3)^3 - 9(3)^2 \stackrel{?}{=} -\dfrac{135}{2}$

3. To find the critical points of $f(x) = 3x^5 - 5x^3 + 1$, we determine where $f'(x) = 0$:

$\left(3x^5 - 5x^3 + 1\right)' = 0$

$15x^4 - 15x^2 = 0$

$15x^2(x^2 - 1) = 0$

$15x^2(x + 1)(x - 1) = 0$

Critical Point: $x = 0, \ x = -1, \ x = 1$

Values

$f(0) = 1$

$f(-1) = 3(-1)^3 - 5(-1)^3 + 1 = 3$

$f(1) = 3(1)^3 - 5(1)^3 + 1 = -1$

5. To find the critical points of $f(x) = \dfrac{x-1}{x^2}$, we determine where $f'(x) = 0$:

$$\left(\frac{x-1}{x^2}\right)' = (x^{-1} - x^{-2})' = 0$$

$$-x^{-2} + 2x^{-3} = 0$$

$$-\frac{1}{x^2} + \frac{2}{x^3} = 0$$

Multiply both sides of the equation by x^3: $\quad -x + 2 = 0$

Critical Point: $\quad x = 2 \quad$ Value: $f(2) = \dfrac{2-1}{2^2} = \dfrac{1}{4}$

7. To find the critical points of $f(x) = \dfrac{x^2}{x-1}$, we determine where $f'(x) = 0$:

$$\left(\frac{x^2}{x-1}\right)' = 0$$

$$\frac{(x-1)(x^2)' - x^2(x-1)}{(x-1)^2} = 0$$

$$\frac{(x-1)(2x) - x^2}{(x-1)^2} = 0$$

$$\frac{x^2 - 2x}{(x-1)^2} = 0$$

$$\frac{x(x-2)}{(x-1)^2} = 0$$

Multiply both sides of the equation by $(x-1)^2$: $\quad x(x-2) = 0$

Critical Points: $\qquad x = 0 \qquad x = 2$

Corrseponding Values: $\quad f(0) = \dfrac{0^2}{0-1} = 0 \qquad f(2) = \dfrac{2^2}{2-1} = 4$

9. Differentiate and factor:

$$f'(x) = (x^3 - 27x)' = 3x^2 - 27 = 3(x^2 - 9) = 2(x+3)(x-3)$$

$$f''(x) = (3x^2 - 27)' = 6x$$

SIGN $f'(x)$:

inc. $+$ c -3 max dec. $-$ c 3 min inc $+$

SIGN $f''(x)$:

con. down $-$ inf. pt. up $+$

Value: $f(0) = 0$

Values:
$f(-3) = (-3)^3 - 27(-3) = 54$
$f(3) = (3)^3 - 27(3) = -54$

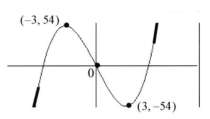

As $x \to \pm\infty$ the graph
resembles that of $y = x^3$

The Graph:

11. Differentiate and factor:

$$f'(x) = (x^4 - 2x^2 + 3)' = 4x^3 - 4x = 4x(x^2 - 1) = 4x(x + 1)(x - 1)$$

$$f''(x) = (4x^3 - 4x)' = 12x^2 - 4 = 4(3x^2 - 1) = 4(\sqrt{3}x + 1)(\sqrt{3}x - 1)$$

SIGN $f'(x)$:

dec.   inc.   dec.   inc.
$-$ $\underset{\underset{\text{min}}{-1}}{\overset{c}{\bullet}}$ $+$ $\underset{\underset{\text{max}}{0}}{\overset{c}{\bullet}}$ $-$ $\underset{\underset{\text{min}}{1}}{\overset{c}{\bullet}}$ $+$

SIGN $f''(x)$:

con. up   down   up
$+$ $\underset{-\frac{1}{\sqrt{3}}}{\bullet}$ $-$ $\underset{\frac{1}{\sqrt{3}}}{\bullet}$ $+$

inf. pt.   inf. pt.

$f(-1) = (-1)^4 - 2(-1)^2 + 3 = 2$
$f(0) = 3$
$f(1) = (1)^4 - 2(1)^2 + 3 = 2$

$f\left(\frac{1}{\sqrt{3}}\right) = \left(\frac{1}{\sqrt{3}}\right)^4 - 2\left(\frac{1}{\sqrt{3}}\right)^2 + 3$

$\qquad = \frac{1}{9} - \frac{2}{9} + 3 = \frac{22}{9}$

$f\left(-\frac{1}{\sqrt{3}}\right) = \frac{22}{9}$

As $x \to \pm\infty$ the graph
resembles that of $y = x^4$

The Graph: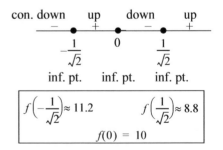

$(-1, 2)$ $(1, 2)$

$\left(-\frac{1}{\sqrt{3}}, \frac{22}{9}\right)$ $\left(-\frac{1}{\sqrt{3}}, \frac{22}{9}\right)$

13. Differentiate and factor:

$$f'(x) = (3x^5 - 5x^3 + 10)' = 15x^4 - 15x^2 = 15x^2(x^2 - 1) = 15x^2(x + 1)(x - 1)$$

$$f''(x) = (15x^4 - 15x^2)' = 60x^3 - 30x = 30x(2x^2 - 1) = 30x(\sqrt{2}x + 1)(\sqrt{2}x - 1)$$

SIGN $f'(x)$:

inc.   dec.   dec.   inc.
$+$ $\underset{\underset{\text{max}}{-1}}{\overset{c}{\bullet}}$ $-$ $\underset{0}{\overset{n}{\bullet}}$ $-$ $\underset{\underset{\text{min}}{1}}{\overset{c}{\bullet}}$ $+$

SIGN$f''(x)$:

con. down   up   down   up
$-$ $\underset{-\frac{1}{\sqrt{2}}}{\bullet}$ $+$ $\underset{0}{\bullet}$ $-$ $\underset{\frac{1}{\sqrt{2}}}{\bullet}$ $+$

inf. pt.   inf. pt.   inf. pt.

$f(-1) = 3(-1)^5 - 5(-1)^3 + 10 = 12$
$f(0) = 10$
$f(1) = 3(1)^5 - 5(1)^3 + 10 = 8$

$f\left(-\frac{1}{\sqrt{2}}\right) \approx 11.2$ $f\left(\frac{1}{\sqrt{2}}\right) \approx 8.8$

$f(0) = 10$

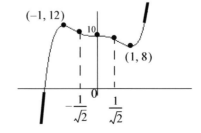

$(-1, 12)$

$(1, 8)$

As $x \to \pm\infty$ the graph
resembles that of $y = x^5$

The Graph:

$-\frac{1}{\sqrt{2}}$ $\frac{1}{\sqrt{2}}$

15. Differentiate and factor:
$$f'(x) = \left(x^4 - \frac{4}{3}x^3 + 1\right)' = 4x^3 - 4x^2 = 4x^2(x-1)$$

$$f''(x) = (4x^3 - 4x^2)' = 12x^2 - 8x = 4x(3x-2)$$

SIGN $f'(x)$:

dec.   dec.   inc.

$\begin{array}{ccc} - & & \\ n & - & c & + \\ \bullet & & \bullet \\ 0 & & 1 \\ & & \text{min.} \end{array}$

$\boxed{\begin{array}{l} f(0) = 1 \\ f(1) = 1 - \frac{4}{3} + 1 = \frac{4}{3} \end{array}}$

SIGN $f''(x)$:

con. up   down   up

$\begin{array}{ccc} + & c & - & c & + \\ \bullet & & \bullet \\ 0 & & \frac{2}{3} \\ \text{inf. pt} & & \text{inf. pt} \end{array}$

$\boxed{\begin{array}{l} f(0) = 1 \\ f\left(\frac{2}{3}\right) = \left(\frac{2}{3}\right)^4 - \frac{4}{3}\left(\frac{2}{3}\right)^3 + 1 = \frac{65}{81} \approx 0.8 \end{array}}$

As $x \to \pm\infty$ the graph resembles that of $y = x^4$

The Graph:

$\left(\frac{2}{3}, \frac{65}{81}\right)$

$\left(1, \frac{2}{3}\right)$

17. Since $f''(x) = (x^2)'' = (2x)' = 2$ is always positive, the graph is concave up everywhere.

positive   positive

Since $f''(x^{2n}) = (x^{2n})'' = (2nx^{2n-1})' = 2n(2n-1)x^{2n-2} = 2n(2n-1)(x^{n-1})^2$ is always positive, the graph is concave up everywhere.

19. $f'(x) = (x^4 - 2x^2 + 1)' = 4x^3 - 4x = 4x(x^2 - 1) = 4x(x+1)(x-1)$

SIGN $f'(x)$:

$\longleftarrow$ decreasding $\quad c \quad$ inc. $\quad c \quad$ dec. $\quad c \quad$ increasing $\longrightarrow$

$\begin{array}{ccccccccc} & - & & + & & - & & + & \\ | & & \bullet & & \bullet & & \bullet & & | & | \\ -2 & & -1 & & 0 & & 1 & & 2 & 3 \\ & & \text{min} & & \text{max} & & \text{min} & & \end{array}$

(a) Over $[-2, 1]$:

$\begin{array}{cccc} & - & + & - \\ & \bullet & \bullet \\ -2 & -1 & 0 & 1 \\ \text{end point max} & \text{min} & \text{max} & \text{min} \end{array}$

$f(-2) = 9 \qquad f(-1) = 0 \qquad f(0) = 1 \qquad f(1) = 0$

Absolute Maximum $\qquad\qquad$ Absolute Minimum
$\qquad\qquad\qquad\qquad\qquad$ (a draw)

(b) Over $[-1, 2]$:

$\begin{array}{cccc} & + & - & + \\ \bullet & \bullet \\ -1 & 0 & 1 & 2 \\ \text{min} & \text{max} & \text{min} & \text{end point max} \end{array}$

$f(-1) = 0 \qquad f(0) = 1 \qquad f(1) = 0 \qquad f(2) = 9$

Absolute Minimim $\qquad\qquad$ Absolute Maximum
(a draw)

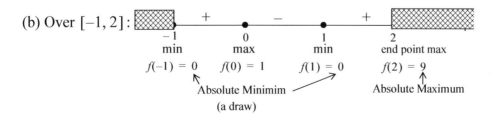

(c) Over $[0, 2]$:

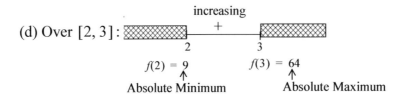

increasing

(d) Over $[2, 3]$:

$f(2) = 9$
↑
Absolute Minimum

$f(3) = 64$
↑
Absolute Maximum

21. For $f(x) = ax^2 + bx + c : f'(x) = 2ax + b$ and $f''(x) = 2a$. It follows that the graph will be concave up if $a > 0$ and concave down if $a < 0$. Moreover, since $f'(x) = 0$ where $2ax + b = 0$, or at $x = -\dfrac{b}{2a}$, the graph will have a minimum at that point if $a > 0$ and a maximum if $a < 0$.

23. $f'(x) = (x^3 - x^2)' = 3x^2 - 2x = x(3x - 2)$ is zero at $x = 0$ and at $x = \dfrac{2}{3}$.

From $f''(x) = (3x^2 - 2x)' = 6x - 2$:

$$f''(0) = -2 < 0 \xrightarrow{\text{Second Derivative Test}} \text{max at } 0$$

$$f''\left(\dfrac{2}{3}\right) = 2 > 0 \xrightarrow{\text{Second Derivative Test}} \text{min at } \dfrac{2}{3}$$

25. $f'(x) = (3x^5 - 5x^3 + 10)' = 15x^4 - 15x^2 = 15x^2(x^2 - 1) = 15x^2(x + 1)(x - 1)$ is zero at $x = 0$ and at $x = \pm 1$.

From $f''(x) = (15x^4 - 15x^2)' = 60x^3 - 30x$:

$$f''(0) = 0 \xrightarrow{\text{Second Derivative Test}} \text{inconclusive}$$

$$f''(1) = 30 > 0 \xrightarrow{\text{Second Der. Test}} \text{min at } 1$$

$$f''(-1) = -30 < 0 \xrightarrow{\text{Second Der. Test}} \text{max at } -1$$

27. (i) If $f(x) = x^3$ then $f'(x) = 3x^2$. SIGN f': $\underset{\substack{0 \\ \text{no max nor min}}}{+ \bullet +}$. $f''(0) = 0$ since $f''(x) = 6x$.

(ii) If $f(x) = x^4$ then $f'(x) = 4x^3$. SIGN f': $\underset{\substack{0 \\ \text{min}}}{- \bullet +}$. $f''(0) = 0$, since $f''(x) = 12x^2$.

(iii) If $f(x) = -x^4$ then $f'(x) = -4x^3$. SIGN f': $\underset{\substack{0 \\ \text{max}}}{+ \bullet -}$. $f''(0) = 0$, since $f''(x) = 12x^2$.

§2.7. Optimization Problems
Page 189

1. For $C(x) = 5000 + 100x$, $R(x) = 200x - \dfrac{x^2}{10}$ we have:

$$P(x) = R(x) - C(x) = \left(200x - \frac{x^2}{10}\right) - (5000 + 100x) = -5000 + 100x - \frac{x^2}{10}$$

So: $P'(x) = 100 - \dfrac{x}{5} = \dfrac{1}{5}(500 - x)$ SIGN $P'(x)$: $\begin{array}{c} + \quad c \quad - \\ \hline \bullet \\ 500 \\ max \end{array}$

From the above we see that maximum profit occurs when 500 units are produced and sold. To see that the maximal marginal profit occurs when marginal revenue equals marginal cost you need only observe that for $P'(x)$ to be zero (horizontal tangent line):

$$P'(x) = R'(x) - C'(x) = 0 \Rightarrow R'(x) = C'(x)$$

$$\text{Note: } \overline{MR}(x) = \overline{MC}(x)$$

3. For $Y(x) = 60 + 16x - 4x^2$ we have:

$$Y'(x) = (60 + 16x - 4x^2)' = 16 - 8x = 8(2 - x)$$ SIGN $Y'(x)$: $\begin{array}{c} + \quad c \quad - \\ \hline \bullet \\ 2 \\ max \end{array}$

Conclusion: Yield will be maximized when 2 quarts of fertilizer per acre is used.

5. For $v(r) = ar^2(r_0 - r)$ we have:

$$\frac{dv}{dr} = (ar_0 r^2 - ar^3)' = 2ar_0 r - 3ar^2 = ar(2r_0 - 3r)$$ SIGN $v'(r)$: $\begin{array}{c} c \quad + \quad c \quad - \\ \boxtimes\!\!\!\!\!\!\bullet\bullet \\ 0 \qquad \frac{2}{3}r_0 \end{array}$

where ar_0 is constant.

Conclusion: Velocity is greatest when $r = \dfrac{2}{3}r_0$.

7.

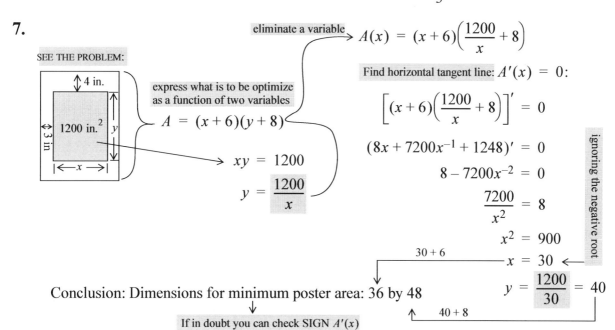

SEE THE PROBLEM:

eliminate a variable $\rightarrow A(x) = (x + 6)\left(\dfrac{1200}{x} + 8\right)$

express what is to be optimize as a function of two variables
$A = (x + 6)(y + 8)$

$xy = 1200$

$y = \dfrac{1200}{x}$

Find horizontal tangent line: $A'(x) = 0$:

$$\left[(x + 6)\left(\frac{1200}{x} + 8\right)\right]' = 0$$

$$(8x + 7200x^{-1} + 1248)' = 0$$

$$8 - 7200x^{-2} = 0$$

$$\frac{7200}{x^2} = 8$$

$$x^2 = 900$$

$$x = 30 \leftarrow$$

ignoring the negative root

$$y = \frac{1200}{30} = 40$$

$30 + 6$

Conclusion: Dimensions for minimum poster area: 36 by 48

$40 + 8$

If in doubt you can check SIGN $A'(x)$

9.

SEE THE PROBLEM

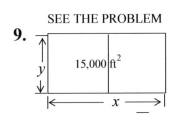

Letting F denote the required feet of fence we have:

$$F = 2x + 3y$$

$$xy = 15000$$

$$y = \frac{15000}{x}$$

divide by x:

$$F(x) = 2x + 3\left(\frac{15000}{x}\right)$$

Finding the horizontal tangent line:

$$F'(x) = 0$$

$$(2x + 45000x^{-1})' = 0$$

$$2 - 45000x^{-2} = 0$$

$$\frac{45000}{x^2} = 2$$

$$x^2 = 22500$$

ignoring the negative root: $x = 150$

$$y = \frac{15000}{150} = 100$$

Conclusion: Dimensions for minimum fence: 150 by 100 ft.

↓

If in doubt you can check SIGN $F'(x)$

11.

SEE THE PROBLEM

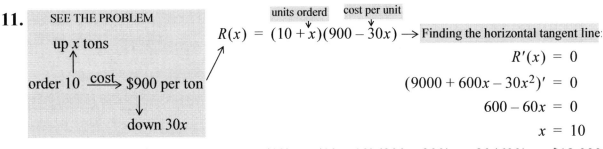

$$R(x) = (10 + x)(900 - 30x) \rightarrow$$

Finding the horizontal tangent line:

$$R'(x) = 0$$

$$(9000 + 600x - 30x^2)' = 0$$

$$600 - 60x = 0$$

$$x = 10$$

Conclusion: Maximum Reveue $= R(10) = (10 + 10)(900 - 300) = 20(600) = \$12{,}000$

↓

If in doubt you can check SIGN $R'(x)$

13.

SEE THE PROBLEM

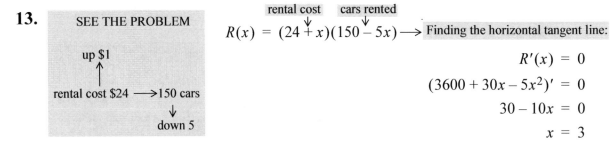

$$R(x) = (24 + x)(150 - 5x) \longrightarrow$$

Finding the horizontal tangent line:

$$R'(x) = 0$$

$$(3600 + 30x - 5x^2)' = 0$$

$$30 - 10x = 0$$

$$x = 3$$

Conclusion: To maximize revenue the agency should charge \$27 per car.

↓

If in doubt you can check SIGN $R'(x)$

15.

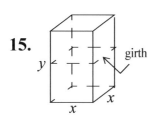

$$V = x^2y \qquad \longrightarrow V(x) = x^2(108 - 4x)$$

girth

To maximize volume let's go all the way up to the allowed girth plus length restriction:

$$4x + y = 108$$
$$y = \boxed{108 - 4x}$$

Finding horizontal tangent lines

$$V'(x) = 0$$
$$(108x^2 - 4x^3)' = 0$$
$$216x - 12x^2 = 0$$
$$12x(18 - x) = 0$$
$$\cancel{x = 0}, \; x = 18 \quad \text{must be positive}$$

Conclusion: For maximum volume the length of a side of the square base should be 18 in.

If in doubt you can check SIGN $V'(x)$

and the height should be $108 - 4 \cdot 18 = 36$ in.

17.

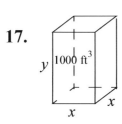

y | 1000 ft³

The surface area, S, is the total area of the sides, top, and bottom of the box:

$$S = 2x^2 + 4xy$$

From $V = x^2y = 1000$

we have: $y = \dfrac{1000}{x^2}$ So:

$$\longrightarrow S(x) = 2x^2 + 4x\frac{1000}{x^2} = 2x^2 + \frac{4000}{x}$$

Finding the horizontal tangent line:

$$S'(x) = 0$$
$$(2x^2 + 4000x^{-1})' = 0$$
$$4x - \frac{4000}{x^2} = 0$$
$$4x^3 - 4000 = 0$$
$$x^3 = 1000$$
$$x = 10$$

Conclusion: Dimensions for minimum surface area: 10 by 10 by 10 ft.

If in doubt you can check SIGN $S'(x)$

19.

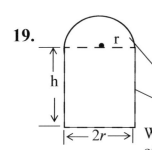

To maximize light is to maximize the total area, of the window:

$$A = 2rh + \frac{1}{2}\pi r^2$$

We are told that the perimeter of the window is 15 feet:

$$2r + 2h + \pi r = 15$$
$$2h = 15 - 2r - \pi r$$
$$h = \frac{15 - 2r - \pi r}{2}$$

$$\longrightarrow A(r) = 2r\frac{15 - 2r - \pi r}{2} + \frac{1}{2}(\pi r^2)$$
$$= 15r - 2r^2 - \frac{\pi}{2}r^2$$

Finding the horizontal tangent line: $A'(r) = 0$

$$\left(15r - 2r^2 - \frac{\pi}{2}r^2\right)' = 0$$
$$15 - 4r - \pi r = 0$$
$$r = \frac{15}{4 + \pi}$$

Dimensions for maximum light: $r = \dfrac{15}{4 + \pi}, h = \dfrac{15 - 2\dfrac{15}{4 + \pi} - \pi\dfrac{15}{4 + \pi}}{2} = \dfrac{15}{4 + \pi} \approx 2.1$ ft.

21.

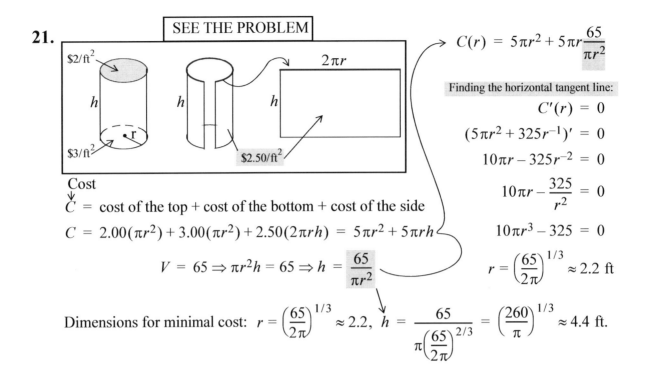

SEE THE PROBLEM

$2/ft^2

h

$3/ft^2

h

$2\pi r$

h

$2.50/ft^2

Cost

C = cost of the top + cost of the bottom + cost of the side

$C = 2.00(\pi r^2) + 3.00(\pi r^2) + 2.50(2\pi rh) = 5\pi r^2 + 5\pi rh$

$V = 65 \Rightarrow \pi r^2 h = 65 \Rightarrow h = \dfrac{65}{\pi r^2}$

$$C(r) = 5\pi r^2 + 5\pi r\dfrac{65}{\pi r^2}$$

Finding the horizontal tangent line:

$$C'(r) = 0$$
$$(5\pi r^2 + 325r^{-1})' = 0$$
$$10\pi r - 325r^{-2} = 0$$
$$10\pi r - \dfrac{325}{r^2} = 0$$
$$10\pi r^3 - 325 = 0$$
$$r = \left(\dfrac{65}{2\pi}\right)^{1/3} \approx 2.2 \text{ ft}$$

Dimensions for minimal cost: $r = \left(\dfrac{65}{2\pi}\right)^{1/3} \approx 2.2, \quad h = \dfrac{65}{\pi\left(\dfrac{65}{2\pi}\right)^{2/3}} = \left(\dfrac{260}{\pi}\right)^{1/3} \approx 4.4 \text{ ft.}$

23. To find the maximum value of:

$$P(x) = R(x) - C(x) = 130x - \sqrt{\dfrac{x^2}{x+100}} - \left(5000 + 100x + \dfrac{x^2}{\sqrt{x+500}}\right)$$

we turn to a calculator:

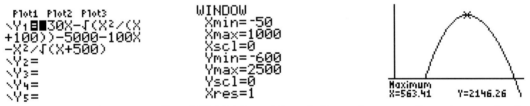

Plot1 Plot2 Plot3
\Y1■30X-√(X²/(X
+100))-5000-100X
-X²/√(X+500)
\Y2=
\Y3=
\Y4=
\Y5=

WINDOW
Xmin=-50
Xmax=1000
Xscl=0
Ymin=-600
Ymax=2500
Yscl=0
Xres=1

Maximum
X=563.41   Y=2146.26

Conclusion: A maximum profit of $2146 is achieved from the manufacturing and sale of 563 units.

25.

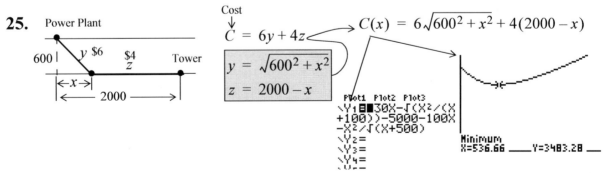

Power Plant

600

y $6

$4 z

Tower

←x→

←——— 2000 ———→

Cost

$C = 6y + 4z$

$y = \sqrt{600^2 + x^2}$

$z = 2000 - x$

$$C(x) = 6\sqrt{600^2 + x^2} + 4(2000 - x)$$

Plot1 Plot2 Plot3
\Y1■30X-√(X²/(X
+100))-5000-100X
-X²/√(X+500)
\Y2=
\Y3=
\Y4=

Minimum
X=536.66 ____Y=3483.28 ___

Conclusion: The most economical route is to lay the pipe in the water to a point that is 536.66 feet from the point directly across the power plant (for a total cost of $3483.28).

27.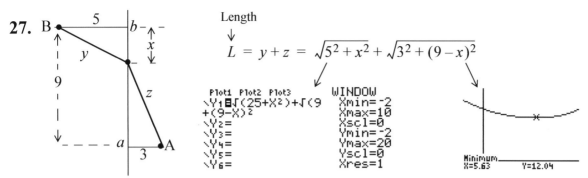

Conclusion: To minimize the cable length the cable from B should connect with a point on the power line that is 5.63 miles south of b (for a total length of 12.04 miles).

§2.8. The Indefinite Integral
Page 199

1. $\int 3\,dx = 3x + C$

3. $\int (3 + 3x)\,dx = 3x + \dfrac{3x^2}{2} + C$

5. $\int (6x^5 + 5x^4)\,dx = 6\dfrac{x^6}{6} + 5\dfrac{x^5}{5} + C = x^6 + x^5 + C$

7. $\int \left(\dfrac{x^4}{5} - \dfrac{3}{x^5} \right) dx = \int \left(\dfrac{x^4}{5} - 3x^{-5} \right) dx = \dfrac{1}{5} \cdot \dfrac{x^5}{5} + 3 \cdot \dfrac{x^{-4}}{-4} + C = \dfrac{x^5}{25} - \dfrac{3}{4x^4} + C$

9. $\int \left(3x^4 - 4x^{-4} + \dfrac{2}{x^5} \right) dx = \int (3x^4 - 4x^{-4} + 2x^{-5})\,dx$

$$= 3 \cdot \dfrac{x^5}{5} - 4 \cdot \dfrac{x^{-3}}{-3} + 2 \cdot \dfrac{x^{-4}}{-4} = \dfrac{3x^5}{5} + \dfrac{4}{3x^3} - \dfrac{1}{2x^4} + C$$

11. $\int x^2(2x - 5)\,dx = \int (2x^3 - 5x^2)\,dx = 2 \cdot \dfrac{x^4}{4} - 5 \cdot \dfrac{x^3}{3} + C = \dfrac{x^4}{2} - \dfrac{5x^3}{3} + C$

13. $\int \dfrac{3x^5 + 2x - 1}{x^4}\,dx = \int (3x + 2x^{-3} - x^{-4})\,dx$

$$= 3 \cdot \dfrac{x^2}{2} + 2 \cdot \dfrac{x^{-2}}{-2} - \dfrac{x^{-3}}{-3} + C = \dfrac{3x^2}{2} - \dfrac{1}{x^2} + \dfrac{1}{3x^3} + C$$

15. $\int \dfrac{(x^4 + x)(x + 1)}{x^4}\,dx = \int \dfrac{x^5 + x^4 + x^2 + x}{x^4}\,dx$

$$= \int (x + 1 + x^{-2} + x^{-3})\,dx = \dfrac{x^2}{2} + x + \dfrac{x^{-1}}{-1} + \dfrac{x^{-2}}{-2} = \dfrac{x^2}{2} + x - \dfrac{1}{x} - \dfrac{1}{2x^2} + C$$

17. We show that $\dfrac{x}{2x+5}$ is an antiderivative of $\dfrac{5}{(2x+5)^2}$:

$$\left[\dfrac{x}{2x+5}\right]' = \dfrac{(2x+5)(x)' - x(2x+5)'}{(2x+5)^2} = \dfrac{2x+5-2x}{(2x+5)^2} = \dfrac{5}{(2x+5)^2}$$

19. We show that $\dfrac{-1}{(2x^2+2)^2}$ is an antiderivative of $\dfrac{x}{(x^2+1)^3}$:

$$\left[\dfrac{-1}{(2x^2+2)^2}\right]' = \dfrac{(2x^2+2)^2(-1)' - (-1)[(2x^2+2)^2]'}{(2x^2+2)^4}$$

$$= \dfrac{(2x^2+2)^2 \cdot 0 + (4x^4+8x^2+4)'}{2^4(x^2+1)^4} = \dfrac{16x^3+16x}{16(x^2+1)^4}$$

$$= \dfrac{16x(x^2+1)}{16(x^2+1)^4} = \dfrac{x}{(x^2+1)^3}$$

Since $f(1) = 5$:

21. $f(x) = \int(3x+5)dx = \dfrac{3x^2}{2} + 5x + \boxed{C}$ $\quad 5 = \dfrac{3(1)^2}{2} + 5(1) + C \Rightarrow \boxed{C = -\dfrac{3}{2}}$

$$f(x) = \dfrac{3x^2}{2} + 5x - \boxed{\dfrac{3}{2}}$$

Since $f(1) = 0$:

$$0 = \dfrac{1^4}{4} + \dfrac{5(1)^2}{2} - 2(1) + C$$

23. $f(x) = \int(x^3+5x-2)dx = \dfrac{x^4}{4} + \dfrac{5x^2}{2} - 2x + \boxed{C}$

$$C = -\dfrac{1}{4} - \dfrac{5}{2} + 2 = \dfrac{-1-10+8}{4} = \boxed{-\dfrac{3}{4}}$$

$$f(x) = \dfrac{x^4}{4} + \dfrac{5x^2}{2} - 2x - \boxed{\dfrac{3}{4}}$$

25. $f(x) = \int\dfrac{3x^2+5x}{x^5}dx = \int(3x^{-3}+5x^{-4})dx$ $\qquad$ Since $f(1) = 2$:

$$= 3\cdot\dfrac{x^{-2}}{-2} + 5\cdot\dfrac{x^{-3}}{-3} + C \qquad 2 = -\dfrac{3}{2} - \dfrac{5}{3} + C$$

$$= -\dfrac{3}{2x^2} - \dfrac{5}{3x^3} + C \qquad C = 2 + \dfrac{3}{2} + \dfrac{5}{3} = \dfrac{12+9+10}{6} = \boxed{\dfrac{31}{6}}$$

$$= -\dfrac{3}{2x^2} - \dfrac{5}{3x^3} + \boxed{\dfrac{31}{6}}$$

27. Integrating the marginal cost function $\overline{MC}(x) = \dfrac{x}{250} + 100$ we have:

$$C(x) = \int\left(\dfrac{x}{250} + 100\right)dx = \dfrac{x^2}{500} + 100x + C \longleftarrow \text{fixed cost}$$

Using the given information that it costs the company $20,000 to produce 100 units [i.e: $C(100) = \$20,000$] we are able to find the fixed cost of the company:

$$20,000 = C(100)$$

$$C(x) = \dfrac{x^2}{500} + 100x + C: \quad 20,000 = \dfrac{100^2}{500} + 100 \cdot 100 + C$$

$$C = 20,000 - 20 - 10,000 = \$9,980$$

Evaluating the cost function $C(x) = \dfrac{x^2}{500} + 100x + 9,980$ at $x = 125$ we have:

$$C(125) = \dfrac{(125)^2}{500} + 100 \cdot 125 + 9,980 = \$22,511.25$$

29. Integrating the marginal revenue function $\overline{MR}(x) = -\dfrac{x}{200} + 75$ we have:

Must be **0**, since $R(0) = 0$ (sell nothing get nothing)

$$R(x) = \int\left(-\dfrac{x}{200} + 75\right)dx = -\dfrac{x^2}{400} + 75x + C$$

When integrating marginal revenue, the constant of integration is **always 0**

Evaluating the revenue function $R(x) = -\dfrac{x^2}{400} + 75x$ at $x = 100$ we have:

$$R(100) = -\dfrac{(100)^2}{400} + 75 \cdot 100 = \$7475$$

31. Integrating the given marginal cost and marginal revenue functions

$$\overline{MC}(x) = -\dfrac{3x}{200} + 25, \text{ and } \overline{MR}(x) = -\dfrac{x}{500} + 75$$

we have:

$$C(x) = \int\left(-\dfrac{3x}{200} + 25\right)dx \qquad R(x) = \int\left(-\dfrac{x}{500} + 75\right)dx$$

$$= -\dfrac{3x^2}{400} + 25x + C \qquad = -\dfrac{x^2}{1000} + 75x + 0$$

Since: $R(0) = 0$

We are given that $C(300) = 10,000$: $10,000 = -\dfrac{3(300)^2}{400} + 25 \cdot 300 + C$

$$C = 10,000 + 675 - 7500 = \$3175$$

So: $C(x) = -\dfrac{3x^2}{400} + 25x + 3175 \quad$ and $\quad R(x) = -\dfrac{x^2}{1000} + 75x$

Evaluating the resulting profit function

$$P(x) = R(x) - C(x) = \left(-\frac{x^2}{1000} + 75x\right) - \left(-\frac{3x^2}{400} + 25x + 3175\right) = \frac{13x^2}{2000} + 50x - 3175$$

at $x = 500$ we have: $P(500) = \frac{13(500)^2}{2000} + 50 \cdot 500 - 3175 = \$23,450$.

33. Since $\overline{MR}(x) = 75 - \frac{x}{1000}$: $R(x) = \int\left(75 - \frac{x}{1000}\right)dx = 75x - \frac{x^2}{2000} + \overset{\text{since } R(0) = 0}{0}$, and

$$\frac{R(x)}{x} = \frac{75x - \frac{x^2}{2000}}{x} = 75 - \frac{x}{2000} \text{ is the demand equation.}$$

35. We are given that the slope of f is x^2; which is to say: $f'(x) = x^2$ (derivatives yield slope of tangent lines). Integrating we have: $f(x) = \int x^2 dx = \frac{x^3}{3} + C$. We are also given that $f(1) = 5$;

bringing us to: $5 = \frac{1^3}{3} + C \Rightarrow C = \frac{14}{3}$. So: $f(x) = \frac{x^3}{3} + \frac{14}{3}$.

$f(x) = \frac{x^3}{3} + C$

37. Turning to Theorem 2.16 we have the velocity and position function for the stone t seconds after it is dropped:

since the stone was dropped, $v_0 = 0$ choosing the ground as referance point, $s_0 = 2304$

$$(1)\ v(t) = -32t \quad \text{and} \quad (2)\ s(t) = -16t^2 + 2304$$

The stone hits the ground when the distance function $s(t)$ is zero. Solving for t in (2) we have:

$$-16t^2 + 2304 = 0$$

we discarded the negative root (t has to be positive)

$$t^2 = \frac{2304}{16} = 144 \Rightarrow t = 12$$

Knowing that the stone will hit the ground in 12 seconds, we evaluate (1) at 12 to find the velocity of the stone on impact: $v(12) = -32 \cdot 12 = -384$ feet per second. The negative sign indicates that the stone is moving in a downward direction on impact (that we knew). By definition, speed is the magnitude of velocity. Hence, the speed at impact is 384 feet per second.

39. Turning to Theorem 2.16 we have the velocity and position function for the object t seconds after it is thrown:

$$(1)\ v(t) = -32t - \overset{v_0}{16} \quad \text{and} \quad (2)\ s(t) = -16t^2 - 16t + \overset{s_0}{96}$$

We first set our sights on determining the impact speed. From (2) we find the time of impact:

$$0 = -16t^2 - 16t + 96 \Rightarrow t^2 + t - 6 = 0 \Rightarrow (t+3)(t-2) = 0 \Rightarrow t = -3, t = 2$$

divide both sides of the equation by -16 t can't be negative

Evaluating (1) at 2 we find the impact velocity: $v(2) = -32 \cdot 2 - 16 = -80$ feet per second. We are told that the object will bounce back up at three-quarter the impact speed; namely, at $\frac{3}{4} \cdot 80 = 60$ feet per second. To find how high up it will go, we turn to "its" velocity and distant functions:

(i) $v(t) = -32t + \overset{v_0}{60}$ and   (ii) $s(t) = -16t^2 + 60t$ $s_0 = 0$ (relative to ground)

When it reaches its maximum height, the velocity is 0. Turning to (i) we have:

$$0 = -32t + 60 \Rightarrow t = \frac{60}{32} = \frac{15}{8} \text{ seconds}$$

Substituting this value of t in (ii) we find the maximum bouncing height:

$$s\left(\frac{15}{8}\right) = -16\left(\frac{15}{8}\right)^2 + 60\left(\frac{15}{8}\right) = \frac{225}{4} = 56.25 \text{ feet}$$

41. Converting the velocity of the car from miles per hour to feet per seconds we have:

these are all 1   and multiplying by 1 does not change anything

$$55\frac{\text{mi}}{\text{hr}} = 55\frac{\text{mi}}{\text{hr}} \cdot \frac{5280 \text{ ft}}{1 \text{ mi}} \cdot \frac{1 \text{ hr}}{60 \text{ min}} \cdot \frac{1 \text{ min}}{60 \text{ sec}} = \frac{55 \cdot 5280}{60 \cdot 60} \frac{\text{ft}}{\text{sec}} = \frac{242}{3} \frac{\text{ft}}{\text{sec}}$$

We are given the constant acceleration: $a = -30\frac{\text{ft}}{\text{sec}^2}$. Integrating we arrive at the velocity and distant function as a function of time (as measured in seconds from when the bakes were first applied):

$$v(t) = \int -30dt = -30t + \overset{v_0}{\frac{242}{3}} \quad \text{and} \quad s(t) = \int\left(-30t + \frac{242}{3}\right)dt = -15t^2 + \frac{242}{3}t$$

$s_0 = 0$ (referece pont: lovstion of car when brakes wer applied)

From $v(t) = -30t + \frac{242}{3}$ we find the time it takes for the car to come to a halt:

$$0 = -30t + \frac{242}{3} \Rightarrow t = \frac{242}{90} \text{ seconds}$$

Substituting this value in the distance function we find the stopping distance:

$$s\left(\frac{242}{90}\right) = -15\left(\frac{242}{90}\right)^2 + \frac{242}{3} \cdot \frac{242}{90} \approx 108 \text{ feet}$$

§2.9. The Definite Integral
Page 212

1. $\int_0^1 3dx = 3x\big|_0^1 = 3\cdot1 - 3\cdot0 = 3$

3. $\int_{-1}^1 (3+3x)dx = \left(3x+3\frac{x^2}{2}\right)\Big|_{-1}^1 = \left(3\cdot1+\frac{3}{2}\cdot1^2\right) - \left[3\cdot(-1)+\frac{3}{2}(-1)^2\right] = \frac{9}{2}+\frac{3}{2} = 6$

5. $\int_1^2 (2x-5)dx = (x^2-5x)\big|_1^2 = (4-10)-(1-5) = -6+4 = -2$

7. $\int_0^1 (x^2+3x-1)dx = \left(\frac{x^3}{3}+\frac{3x^2}{2}-x\right)\Big|_0^1 = \left(\frac{1}{3}+\frac{3}{2}-1\right)-(0) = \frac{5}{6}$

9. $\int_{-1}^1 x^3dx = \frac{x^4}{4}\Big|_{-1}^1 = \frac{1}{4}-\frac{(-1)^4}{4} = \frac{1}{4}-\frac{1}{4} = 0$

11. $\int_{-1}^2 \frac{x^3-4x+1}{5}dx = \frac{1}{5}\int_{-1}^2 (x^3-4x+1)dx = \frac{1}{5}\cdot\left(\frac{x^4}{4}-2x^2+x\right)\Big|_{-1}^2$

$= \frac{1}{5}\left[\left(\frac{16}{4}-8+2\right)-\left(\frac{1}{4}-2-1\right)\right] = \frac{3}{20}$

13. $\int_1^2 x(3x-2)dx = \int_1^2 (3x^2-2x)dx = (x^3-x^2)\big|_1^2 = (8-4)-(1-1) = 4$

15. $\int_0^{-1} (3x-1)(x-1)dx = \int_0^{-1} (3x^2-4x+1)dx = (x^3-2x^2+x)\big|_0^{-1}$

$= -1-2-1-(0) = -4$

17. $\int_1^2 \frac{(x^4+x)(x+1)}{x^4}dx = \int_1^2 \frac{x^5+x^4+x^2+x}{x^4}dx$

$= \int_1^2 (x+1+x^{-2}+x^{-3})dx = \left(\frac{x^2}{2}+x-x^{-1}-\frac{x^{-2}}{2}\right)\Big|_1^2$

$= \left(\frac{x^2}{2}+x-\frac{1}{x}-\frac{1}{2x^2}\right)\Big|_1^2$

$= \left(2+2-\frac{1}{2}-\frac{1}{8}\right)-\left(\frac{1}{2}+1-1-\frac{1}{2}\right) = \frac{27}{8}$

19. Area $= \int_0^1 x^3 dx = \left.\frac{x^4}{4}\right|_0^1 = \frac{1}{4} - (0) = \frac{1}{4}$

21. Area $= \int_{-1}^1 (-x^3 + x^2)dx = \left.\left(-\frac{x^4}{4} + \frac{x^3}{3}\right)\right|_{-1}^1 = \left(-\frac{1}{4} + \frac{1}{3}\right) - \left(-\frac{1}{4} - \frac{1}{3}\right) = \frac{2}{3}$

23. Appealing directly to Theorem 2.18 we have: $T'(x) = \dfrac{1}{x^2 + 4}$.

25. First step: $T(x) = \int_x^5 \sqrt{3t^4 + 1}\, dt = -\int_5^x \sqrt{3t^4 + 1}\, dt$. Second step: $T'(x) = -\sqrt{3x^4 + 1}$.
$\qquad\qquad\qquad\qquad\qquad\qquad\quad \uparrow \qquad\qquad\qquad\qquad\qquad\qquad\qquad\qquad\qquad \uparrow$
$\qquad\qquad\qquad\qquad\qquad\quad$ Definiton 2.12 $\qquad\qquad\qquad\qquad\qquad\qquad\quad$ Theorem 2.18

27. Using the Principal Theorem of Calculus: $T'(x) = x^2 + x$.
Without the Principal Theorem of Calculus:

First: $T(x) = \int_1^x (t^2 + t)dt = \left.\left(\frac{t^3}{3} + \frac{t^2}{2}\right)\right|_1^x = \left(\frac{x^3}{3} + \frac{x^2}{2}\right) - \left(\frac{1}{3} + \frac{1}{2}\right) = \frac{x^3}{3} + \frac{x^2}{2} - \frac{5}{6}$

Then: $T'(x) = \left(\frac{x^3}{3} + \frac{x^2}{2} - \frac{5}{6}\right)' = x^2 + x$

29. Noting that the price continues to increase throughout the course of the month we have:

Increase $= \int_0^{31} (0.06t + 0.001t^2)dt = \left.\left(0.06\frac{t^2}{2} + 0.001\frac{t^3}{3}\right)\right|_0^{31} = 0.06 \cdot \frac{31^2}{2} + 0.001 \cdot \frac{31^3}{3} \approx \$0.39$

31. Letting T denote the minutes required for all of 360 cubic inch block to melt we have:

$$\int_0^T \frac{t}{5}dt = 360$$

$$\left.\frac{t^2}{10}\right|_0^T = 360 \Rightarrow \frac{T^2}{10} = 360 \Rightarrow T^2 = 3600 \Rightarrow T = 60 \text{ minutes}$$

33. Letting T denote the number of days needed for the daily customers to increase by 200 we have:

$$\int_0^T \frac{t}{100}dt = 200$$

$$\left.\frac{t^2}{200}\right|_0^T = 200 \Rightarrow \frac{T^2}{200} = 200 \Rightarrow T^2 = 40,000 \Rightarrow T = 200 \text{ days}$$

35. Here is how much the machine will depreciate over 10 years:

$$\int_0^{10} 250(15-x)dx = \left(3750x - 250\frac{x^2}{2}\right)\Big|_0^{10} = 3750 \cdot 10 - 125 \cdot 10^2 = \$25,000$$

We are given that the original value of the machine was \$28,000. So, after 10 years, the value of the machine is $\$(28,000 - 25,000) = \$3000$.

37. We determine the two year increase in earnings, I_2 and I_4, for the \$1000 and the \$2000 machines, respectively:

$$I_2 = \int_0^{24} (190 + 2t)dt \qquad\qquad I_2 = \int_0^{24} (200 + 4t)dt$$

$$= (190t + t^2)\Big|_0^{24} \qquad\qquad = (200t + 2t^2)\Big|_0^{24}$$

$$= 190 \cdot 24 + 24^2 = \$5136 \qquad = 200 \cdot 24 + 2 \cdot 24^2 = \$5952$$

Subtracting the two-year rental cost from the above increased earnings we conclude that the company should rent the less expensive machine:

$$\$5136 - 2(\$1000) = \$3136 \text{ verses } \$5952 - 2(\$2000) = \$1952$$

39. Recalling that $\overline{MR}(x) = R'(x)$ (the rate of change of the revenue function) we apply Theorem 2.20 we determine the net-change in revenue from the sale of 100 through 200 units:

$$\int_{100}^{200} \left(-\frac{x}{500} + 50\right)dx = \left(-\frac{x^2}{1000} + 50x\right)\Big|_{100}^{200}$$

$$= \left(-\frac{(200)^2}{1000} + 50 \cdot 200\right) - \left(-\frac{(100)^2}{1000} + 50 \cdot 100\right) = \$4970$$

41. (a) Recalling that velocity is the rate of change of position with respect to time we apply Theorem 2.20 we determine the net change in the particle's position during the first 4 seconds:

$$\int_0^4 (t^2 - 3t + 2)dt = \left(\frac{t^3}{3} - \frac{3t^2}{2} + 2t\right)\Big|_0^4 = \frac{4^3}{3} - \frac{3 \cdot 16}{2} + 8 = \frac{16}{3} \text{ meters}$$

(b) A consideration of SIGN $v(t) = (t-2)(t-1)$: reveals that the particle starts off moving in an easterly direction until it comes to a stop at $t = 1$. It then moves in a westerly direction until it again comes to a stop at $t = 2$; after which it continues to move in an easterly direction — something like this:

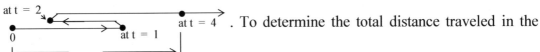

. To determine the total distance traveled in the first four seconds we have to add the distance traveled in the first second to that traveled in the next second, to that traveled in the next two seconds. Finding those three distances we have:

First second: $\int_0^1 (t^2 - 3t + 2)dt = \left(\dfrac{t^3}{3} - \dfrac{3t^2}{2} + 2t\right)\Big|_0^1 = \dfrac{1}{3} - \dfrac{3}{2} + 2 = \dfrac{5}{6}$ m

moving in a westerly direction

Second second: $\int_1^2 (t^2 - 3t + 2)dt = \left(\dfrac{t^3}{3} - \dfrac{3t^2}{2} + 2t\right)\Big|_1^2 = \left(\dfrac{8}{3} - 6 + 4\right) - \left(\dfrac{1}{3} - \dfrac{3}{2} + 2\right) = -\dfrac{1}{6}$

Distance traveled $= \dfrac{1}{6}$ m

Next 2 seconds: $\int_2^4 (t^2 - 3t + 2)dt = \left(\dfrac{t^3}{3} - \dfrac{3t^2}{2} + 2t\right)\Big|_2^4 = \left(\dfrac{4^3}{3} - 24 + 8\right) - \left(\dfrac{8}{3} - 6 + 4\right) = \dfrac{14}{3}$ m

Conclusion: Distance traveled during first four seconds is $\dfrac{5}{6} + \dfrac{1}{6} + \dfrac{14}{3} = \dfrac{17}{3}$ meters.

43. If you choose a rational number in each Δx of any partition of $[0, 1]$, then

$\displaystyle\sum_0^1 f(x)\Delta x = \sum_0^1 1\Delta x = 1$. On the other hand, if you choose an irrational number in each

Δx, then $\displaystyle\sum_0^1 f(x)\Delta x = \sum_0^1 0\Delta x = 0$. It follows that $\displaystyle\lim_{\Delta x \to 0}\sum_a^b f(x)\Delta x$ does not exist.

45.

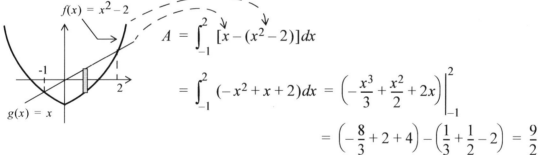

$f(x) = x^2 - 2$

$g(x) = x$

$A = \int_{-1}^2 [x - (x^2 - 2)]dx$

$= \int_{-1}^2 (-x^2 + x + 2)dx = \left(-\dfrac{x^3}{3} + \dfrac{x^2}{2} + 2x\right)\Big|_{-1}^2$

$= \left(-\dfrac{8}{3} + 2 + 4\right) - \left(\dfrac{1}{3} + \dfrac{1}{2} - 2\right) = \dfrac{9}{2}$

47.

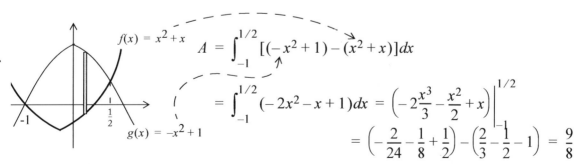

$f(x) = x^2 + x$

$g(x) = -x^2 + 1$

$A = \int_{-1}^{1/2} [(-x^2 + 1) - (x^2 + x)]dx$

$= \int_{-1}^{1/2} (-2x^2 - x + 1)dx = \left(-2\dfrac{x^3}{3} - \dfrac{x^2}{2} + x\right)\Big|_{-1}^{1/2}$

$= \left(-\dfrac{2}{24} - \dfrac{1}{8} + \dfrac{1}{2}\right) - \left(\dfrac{2}{3} - \dfrac{1}{2} - 1\right) = \dfrac{9}{8}$

49.

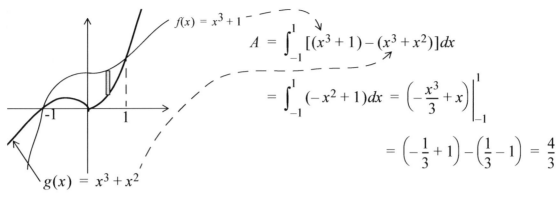

$$A = \int_{-1}^{1} [(x^3 + 1) - (x^3 + x^2)]dx$$

$$= \int_{-1}^{1} (-x^2 + 1)dx = \left(-\frac{x^3}{3} + x\right)\Big|_{-1}^{1}$$

$$= \left(-\frac{1}{3} + 1\right) - \left(\frac{1}{3} - 1\right) = \frac{4}{3}$$

51.

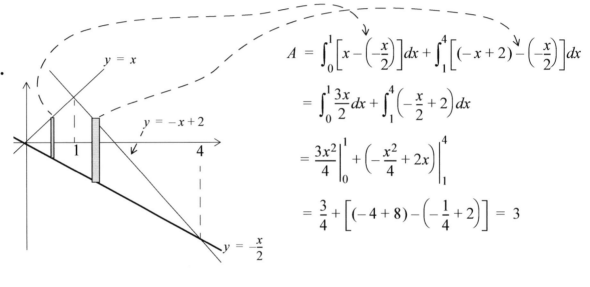

$$A = \int_{0}^{1} \left[x - \left(-\frac{x}{2}\right)\right]dx + \int_{1}^{4} \left[(-x + 2) - \left(-\frac{x}{2}\right)\right]dx$$

$$= \int_{0}^{1} \frac{3x}{2}dx + \int_{1}^{4} \left(-\frac{x}{2} + 2\right)dx$$

$$= \frac{3x^2}{4}\Big|_{0}^{1} + \left(-\frac{x^2}{4} + 2x\right)\Big|_{1}^{4}$$

$$= \frac{3}{4} + \left[(-4 + 8) - \left(-\frac{1}{4} + 2\right)\right] = 3$$

53.

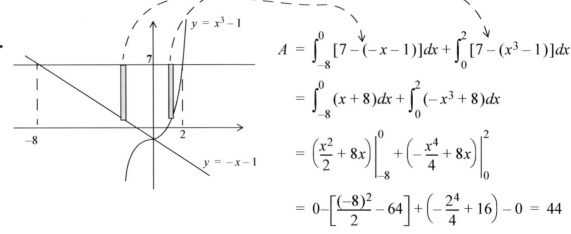

$$A = \int_{-8}^{0} [7 - (-x - 1)]dx + \int_{0}^{2} [7 - (x^3 - 1)]dx$$

$$= \int_{-8}^{0} (x + 8)dx + \int_{0}^{2} (-x^3 + 8)dx$$

$$= \left(\frac{x^2}{2} + 8x\right)\Big|_{-8}^{0} + \left(-\frac{x^4}{4} + 8x\right)\Big|_{0}^{2}$$

$$= 0 - \left[\frac{(-8)^2}{2} - 64\right] + \left(-\frac{2^4}{4} + 16\right) - 0 = 44$$

§Review Exercises
Page 222

1. $-5^2 = -(5 \cdot 5) = -25$ (Note the difference: $(-5)^2 = (-5)(-5) = 25$)

2. $(-5)^{-2} = \dfrac{1}{(-5)^2} = \dfrac{1}{25}$

3. $\left(\dfrac{2}{3}\right)^{-2} = \dfrac{1}{\left(\dfrac{2}{3}\right)^2} = \dfrac{1}{\dfrac{4}{9}} = \dfrac{\dfrac{1}{1}}{\dfrac{4}{9}} = \dfrac{1}{1}\cdot\dfrac{9}{4} = \dfrac{9}{4}$ invert and multiply

4. $\dfrac{2^{-1}+2}{9\cdot 3^{-1}} = \dfrac{\dfrac{1}{2}+2}{9\cdot\dfrac{1}{3}} = \dfrac{\dfrac{1}{2}+\dfrac{4}{2}}{3} = \dfrac{\dfrac{5}{2}}{3} = \dfrac{\dfrac{5}{2}}{\dfrac{3}{1}} = \dfrac{5}{2}\cdot\dfrac{1}{3} = \dfrac{5}{6}$ invert ant multiply

5. $\dfrac{(ab)^2}{a^3(-b)^3} = \dfrac{a^2b^2}{-a^3b^3} = \dfrac{1}{-ab} = -\dfrac{1}{ab}$

6. $\dfrac{(-ab)^2a^3}{(2ab)^3} = \dfrac{a^2b^2a^3}{8a^3b^3} = \dfrac{a^2}{8b}$ "gone"

7. $\dfrac{(ab)^2}{a^3-a^2b} = \dfrac{a^2b^2}{a^2(a-b)} = \dfrac{b^2}{a-b}$ Remember: You can **only** cancel a **common factor**: $\dfrac{ac}{bc} = \dfrac{a}{b}$ (for $c \neq 0$)

8. $\dfrac{(ab)^2+a^2}{(-ab)^2} = \dfrac{a^2b^2+a^2}{a^2b^2} = \dfrac{a^2(b^2+1)}{a^2b^2} = \dfrac{b^2+1}{b^2}$ (Don't even think about canceling the b^2)

9. $x^4-16 = (x^2+4)(x^2-4) = (x^2+4)(x+2)(x-2)$

10. $x^3-9x = x(x^2-9) = x(x+3)(x-3)$

11. $4x^4-9 = (2x^2+3)(2x^2-3) = (2x^2+3)(\sqrt{2}x+\sqrt{3})(\sqrt{2}x-\sqrt{3})$

12. $6x^2-15x+6 = 3(2x^2-5x+2) = 3(x-2)(2x-1)$

13. $4x^2-16x+16 = 4(x^2-4x+4) = 4(x-2)(x-2) = 4(x-2)^2$

14. $4x^3-16x^2+16x = 4x(x^2-4x+4) = 4x(x-2)(x-2) = 4x(x-2)^2$

15. Slope 5 and y-intercept 4: $y = 5x+4$.

16. Slope: $m = \dfrac{9-7}{3-2} = \dfrac{2}{1} = 2$. So: $y = 2x + b$. Since $(x, y) = (\mathbf{3}, \mathbf{9})$ is on the line [could have chosen (2,7)]: $\mathbf{9} = 2 \cdot \mathbf{3} + b \Rightarrow \mathbf{b} = 9 - 6 = 3$. Equation: $y = 2x + 3$.

17.
$$-2x + 19 = 5x - 3$$
$$-2x - 5x = -3 - 19$$
$$-7x = -21$$
$$x = 3$$

18.
$$5x + 4 = -4x + 1$$
$$5x + 4x = 1 - 4$$
$$9x = -3$$
$$x = -\frac{1}{3}$$

19.
$$x^2 - 2x - 15 = 0$$
$$(x - 5)(x + 3) = 0$$
$$x = 5, x = -3$$

20.
$$x^2 + 4x + 3 = 0$$
$$(x + 3)(x + 1) = 0$$
$$x = -3, x = 1$$

21.
$$7x^2 - 10x - 8 = 0$$
$$(7x + 4)(x - 2) = 0$$
$$x = -\frac{4}{7}, x = 2$$

22.
$$2x^5 - 7x^4 + 3x^3 = 0$$
$$x^3(2x^2 - 7x + 3) = 0$$
$$x^3(2x - 1)(x - 3) = 0$$
$$x = 0, x = \frac{1}{2}, x = 3$$

23.
$$x + 4 < 3x + 5$$
$$x - 3x < 5 - 4$$
$$-2x < 1$$
$$x > -\frac{1}{2} \quad \text{or:} \quad \left(-\frac{1}{2}, \infty\right)$$

24.
$$-2x + 1 \geq 5x + 9$$
$$-2x - 5x \geq 9 - 1$$
$$-7x \geq 8$$
$$x \leq -\frac{8}{7} \quad \text{or:} \quad \left(-\infty, -\frac{8}{7}\right]$$

25. $x(5x - 3)(x + 2)(x - 7) < 0$

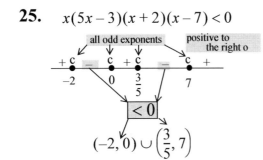

$$(-2, 0) \cup \left(\frac{3}{5}, 7\right)$$

26. $x^4 - x^3 - 6x^2 \geq 0 \Rightarrow x^2(x^2 - x - 6) \geq 0$

$$x^2(x - 3)(x + 2) \geq 0$$

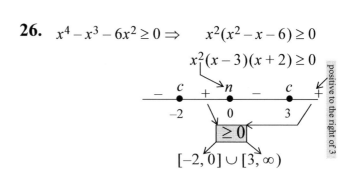

$$[-2, 0] \cup [3, \infty)$$

27. $\displaystyle\lim_{x \to 2} \frac{x - 2}{x + 2} = \frac{0}{4} = 0$

denominator is not 0 at $x = 2$

28. $\displaystyle\lim_{x \to 3} \frac{x^2 - 9}{2x^2 - 5x - 3} = \lim_{x \to 3} \frac{(x + 3)(x - 3)}{(2x + 1)(x - 3)} = \lim_{x \to 3} \frac{x + 3}{2x + 1} = \frac{6}{7}$

29. $\displaystyle\lim_{x \to -4} \frac{2x^2 + 5x - 12}{x^2 + x - 12} = \lim_{x \to -4} \frac{(2x - 3)(x + 4)}{(x - 3)(x + 4)} = \lim_{x \to -4} \frac{2x - 3}{x - 3} = \frac{-11}{-7} = \frac{11}{7}$

$$f(x) = 3x + 10$$

30. $f'(1) = \lim_{h \to 0} \dfrac{f(1+h) - f(1)}{h} = \lim_{h \to 0} \dfrac{3(1+h) + 10 - (3 \cdot 1 + 10)}{h}$

$$= \lim_{h \to 0} \frac{3 + 3h + 10 - 13}{h} = \lim_{h \to 0} \frac{3h}{h} = \lim_{h \to 0} 3 = 3$$

$$f(x) = 2x^2 + x - 1$$

31. $f'(1) = \lim_{h \to 0} \dfrac{f(1+h) - f(1)}{h} = \lim_{h \to 0} \dfrac{2(1+h)^2 + (1+h) - 1 - (2 \cdot 1^2 + 1 - 1)}{h}$

$$= \lim_{h \to 0} \frac{2(1 + 2h + h^2) + h - 2}{h}$$

$$= \lim_{h \to 0} \frac{2 + 4h + 2h^2 + h - 2}{h} = \lim_{h \to 0} \frac{5h + 2h^2}{h}$$

$$= \lim_{h \to 0} \frac{h(5 + 2h)}{h} = \lim_{h \to 0} (5 + 2h) = 5$$

$$f(x) = \frac{x}{2x + 1}$$

32. $f'(1) = \lim_{h \to 0} \dfrac{f(1+h) - f(1)}{h} = \lim_{h \to 0} \dfrac{\dfrac{1+h}{2(1+h) + 1} - \dfrac{1}{2 \cdot 1 + 1}}{h}$

$$= \lim_{h \to 0} \frac{\dfrac{1+h}{3 + 2h} - \dfrac{1}{3}}{h} = \lim_{h \to 0} \frac{\dfrac{3(1+h) - 1(3 + 2h)}{3(3 + 2h)}}{\dfrac{h}{1}}$$

$$= \lim_{h \to 0} \frac{3 + 3h - 3 - 2h}{3(3 + 2h)h}$$

$$= \lim_{h \to 0} \frac{h}{3(3 + h)h} = \lim_{h \to 0} \frac{1}{3(3 + h)} = \frac{1}{9}$$

$$f(x) = 2x + 1$$

33. $f'(x) = \lim_{h \to 0} \dfrac{f(x+h) - f(x)}{h} = \lim_{h \to 0} \dfrac{2(x+h) + 1 - (2x + 1)}{h}$

$$= \lim_{h \to 0} \frac{2x + 2h + 1 - 2x - 1}{h} = \lim_{h \to 0} \frac{2h}{h} = \lim_{h \to 0} 2 = 2$$

34. $f'(x) = \lim_{h \to 0} \dfrac{f(x+h)-f(x)}{h} = \lim_{h \to 0} \dfrac{-(x+h)^2 + 3(x+h) + 2 - (-x^2 + 3x + 2)}{h}$

$f(x) = -x^2 + 3x + 2$

$= \lim_{h \to 0} \dfrac{-(x^2 + 2xh + h^2) + 3x + 3h + 2 + x^2 - 3x - 2}{h}$

$= \lim_{h \to 0} \dfrac{-x^2 - 2xh - h^2 + 3h + x^2}{h} = \lim_{h \to 0} \dfrac{h(-2x - h + 3)}{h}$

$= \lim_{h \to 0} (-2x - h + 3) = -2x + 3$

35. $f'(x) = \lim_{h \to 0} \dfrac{f(x+h)-f(x)}{h} = \lim_{h \to 0} \dfrac{\frac{(x+h)+1}{2(x+h)} - \frac{x+1}{2x}}{h}$

$f(x) = \dfrac{x+1}{2x}$

$= \lim_{h \to 0} \dfrac{\frac{x(x+h+1) - (x+h)(x+1)}{2x(x+h)}}{h}$

$= \lim_{h \to 0} \dfrac{x^2 + xh + x - (x^2 + x + xh + h)}{2x(x+h)h}$

$= \lim_{h \to 0} \dfrac{-h}{2x(x+h)h} = \lim_{h \to 0} \dfrac{-1}{2x(x+h)} = -\dfrac{1}{2x^2}$

36. $(4x^3 + 2x^2 - x - 1)' = 12x^2 + 4x - 1$

37. $(-3x^2 + 2x - 5x^{-1} - 1)' = -6x + 2 + 5x^{-2} = -6x + 2 + \dfrac{5}{x^2}$

38. $[(x^2 + 1)(x-3)]' = (x^3 - 3x^2 + x - 3)' = 3x^2 - 6x + 1$

39. $\left[\dfrac{3x^5 - 2x + 1}{x^3}\right]' = (3x^2 - 2x^{-2} + x^{-3})' = 6x + 4x^{-3} - 3x^{-4} = 6x + \dfrac{4}{x^3} - \dfrac{3}{x^4}$

40. $\left(\dfrac{x^2}{x-3}\right)' \overset{\text{Theorem 2.10, page 154}}{=} \dfrac{(x-3)(x^2)' - x^2(x-3)'}{(x-3)^2} = \dfrac{(x-3)(2x) - x^2(1)}{(x-3)^2} = \dfrac{x^2 - 6x}{(x-3)^2}$

41. $\left(\dfrac{x^2+x}{2x-1}\right)' \overset{\text{Theorem 2.10, page 154}}{=} \dfrac{(2x-1)(x^2+x)'-(x^2+x)(2x-1)'}{(2x-1)^2}$

$$= \frac{(2x-1)(2x+1)-(x^2+x)\cdot 2}{(2x-1)^2} = \frac{4x^2-1-2x^2-2x}{(2x-1)^2} = \frac{2x^2-2x-1}{(2x-1)^2}$$

42. For $f(x) = 2x^3 - 3x$, $f'(x) = 6x^2 - 3$. The slope of tangent line at $x = 1$ is therefore $f'(1) = 6\cdot 1^2 - 3 = \mathbf{3}$. To find the value of b in the equation $y = \mathbf{3}x + b$ of the tangent line we use the fact that the point $(1, f(1)) = (1, 2\cdot 1^3 - 3\cdot 1) = (1, -1)$ lies on the tangent line:

$$y = 3x + b$$
$$-1 = 3\cdot 1 + b \Rightarrow b = -4 \qquad \text{Tangent line: } y = 3x - 4$$

43. For $f(x) = 4x^3 + 2x^2 - x - 1$, $f'(x) = 12x^2 + 4x - 1$. The slope of tangent line at $x = 0$ is therefore $f'(0) = -1$. To find the value of b in the equation $y = -1x + b$ of the tangent line we use the fact that the point $(0, f(0)) = (0, -1)$ lies on the tangent line:

$$y = -1x + b$$
$$-1 = -1\cdot 0 + b \Rightarrow b = -1 \qquad \text{Tangent line: } y = -x - 1$$

44. To say that the tangent line is horizontal is to say that it's slope is 0; which is to say that:

$$f'(x) = (3x^4 - 2x^3 + 100)' = 0$$
$$12x^3 - 6x^2 = 0$$
$$6x^2(2x - 1) = 0$$
$$x = 0, x = \frac{1}{2}$$

45. To say that the tangent line is horizontal is to say that it's slope is 0; which is to say that:

$$f'(x) = \left(x^3 - \frac{x^2}{2} - 2x + 1\right)' = 0$$
$$3x^2 - x - 2 = 0$$
$$(3x + 2)(x - 1) = 0$$
$$x = -\frac{2}{3}, x = 1$$

46. Recalling that the instantaneous rate of change of $f(x) = 3x^2 - 2x + 5$ at $x = 2$ is just another name for $f'(2)$, we have:

$$f'(x) = (3x^2 - 2x + 5)' = 6x - 2 \Rightarrow f'(2) = 10$$

47. Recalling that the instantaneous rate of change of $f(x) = \dfrac{x^2+1}{2x}$ at $x = -1$ is just another

name for $f'(-1)$, we have:

$$f'(x) = \left(\frac{x^2+1}{2x}\right)' = \left(\frac{x}{2}+\frac{x^{-1}}{2}\right)' = \frac{1}{2} - \frac{x^{-2}}{2} = \frac{1}{2} - \frac{1}{2x^2} \Rightarrow f'(-1) = \frac{1}{2} - \frac{1}{2(-1)^2} = 0$$

48. Recalling that marginal profit is simply the derivative of the profit function we have:

$$\overset{\overline{MP}(x)}{\overbrace{P(x) = R(x) - C(x) \Rightarrow P'(x)}} = R'(x) - C'(x)$$
$$= (51x)' - (550 + 26x - 0.05x^2)'$$
$$= 51 - [26 - (0.05)(2x)] = 25 + 0.1x$$

In particular: $\overline{MP}(100) = P'(100) = 25 + 0.1(100) = \$35$; which predicts that the company will realize an approximate profit of \$35 from the sale of the 101th unit.

49. Recalling that marginal profit is simply the derivative of the profit function we have:

$$\overset{\overline{MP}(x)}{\overbrace{P(x) = R(x) - C(x) \Rightarrow P'(x)}} = R'(x) - C'(x)$$
$$= \left(50x - \frac{x^2}{200}\right)' - \left(1000 + 37x + \frac{x^2}{200}\right)'$$
$$= 50 - \frac{x}{100} - \left(37 + \frac{x}{100}\right) = 13 - \frac{x}{50}$$

In particular: $\overline{MP}(100) = P'(100) = 13 - \dfrac{100}{50} = \$11$; which predicts that the company will realize an approximate profit of \$11 from the sale of the 101th unit.

50. To get the cost function we integrate the given marginal cost function:

$$C(x) = \int\left(-\frac{x}{50} + 50\right)dx = -\frac{x^2}{100} + 50x + C \leftarrow \begin{array}{c}\text{constant of integration}\\ \text{(fixed cost of the compn)}\end{array}$$

The given information, $C(100) = \$10{,}000$, enables us to determine the fixed cost:

$$C(x) = -\frac{x^2}{100} + 50x + C \Rightarrow \overset{C(100)\,=\,10{,}000}{10{,}000} = -\frac{100^2}{100} + 50 \cdot 100 + C$$
$$C = 10{,}000 + \frac{100^2}{100} - 50 \cdot 100 = \$5100$$

Turning to the cost function $C(x) = -\dfrac{x^2}{100} + 50x + 5100$ we have:

$$C(150) = -\frac{150^2}{100} + 50 \cdot 150 + 5100 = \$12{,}375$$

51. To get the revenue function we integrate the given marginal revenue function:

$$R(x) = \int\left(-\frac{x}{1000} + 75\right)dx = -\frac{x^2}{2000} + 75x + 0 \leftarrow$$

> the constant of integration must be zero since $R(0) = 0$ (sell nothing make nothing) — see Example 2.39.

In particular: $R(500) = -\frac{500^2}{2000} + 75 \cdot 500 = \$37,375$.

52. From the given information $\overline{MC} = -\frac{x}{100} + 45$, and $\overline{MR} = -\frac{x}{500} + 80$ we have:

$$C(x) = -\frac{x^2}{200} + 45x + C$$

> the given fixed cost of the company

$$C(x) = -\frac{x^2}{200} + 45x + 5000$$

$$R(x) = -\frac{x^2}{1000} + 80x + 0$$

> sell nothing make nothing

In particular:

$$P(500) = R(500) - C(500) = \left(-\frac{500^2}{1000} + 80 \cdot 500\right) - \left(-\frac{500^2}{200} + 45 \cdot 500 + 5000\right) = \$13,500$$

53. $f'(x) = (x^4 + x^2 - 2)'$

$= 4x^3 + 2x$

$= 2x(2x^2 + 1)$

dec c inc

Value: $f(0) = -2$

as $x \to \pm\infty$ the graph of f resembes that of x^4

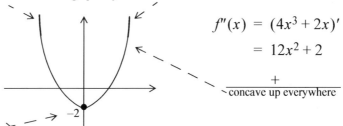

$f''(x) = (4x^3 + 2x)'$

$= 12x^2 + 2$

$+$
concave up everywhere

54. $f'(x) = (x^4 - 4x^3)'$

$= 4x^3 - 12x^2$

$= 4x^2(x - 3)$

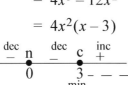

dec dec inc
 n   c

min

Vales: $f(0) = 0$

$f(3) = 3^4 - 4 \cdot 3^3 = -27$

as $x \to \pm\infty$ the graph of f resembes that of x^4

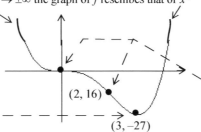

(2, 16)

(3, −27)

$f''(x) = (4x^3 - 12x^2)'$

$= 12x^2 - 24x$

$= 12x(x - 2)$

up c down c up
 +   −   +
        0

Vales: $f(0) = 0$

$f(2) = 2^4 - 4 \cdot 2^3 = -16$

55.
$$f'(x) = \left(x^3 + \frac{x^2}{2} - 2x + 1\right)'$$
$$= 3x^2 + x - 2$$
$$= (3x - 2)(x + 1)$$

as $x \to \pm\infty$ the graph of f resembles that of x^3

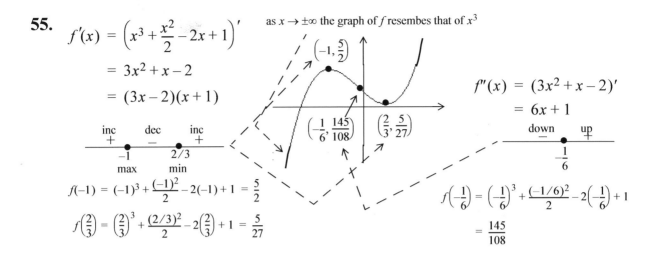

$$f''(x) = (3x^2 + x - 2)'$$
$$= 6x + 1$$

inc   dec   inc
 +   −   +
——●————●————
  −1   2/3
 max   min

$$f(-1) = (-1)^3 + \frac{(-1)^2}{2} - 2(-1) + 1 = \frac{5}{2}$$

$$f\left(\frac{2}{3}\right) = \left(\frac{2}{3}\right)^3 + \frac{(2/3)^2}{2} - 2\left(\frac{2}{3}\right) + 1 = \frac{5}{27}$$

down   up
 −   +
————●————
   −\frac{1}{6}

$$f\left(-\frac{1}{6}\right) = \left(-\frac{1}{6}\right)^3 + \frac{(-1/6)^2}{2} - 2\left(-\frac{1}{6}\right) + 1$$
$$= \frac{145}{108}$$

56.
$$f'(x) = (3x^5 - 25x^3 + 60x)'$$
$$= 15x^4 - 75x^2 + 60$$
$$= 15(x^4 - 5x^2 + 4)$$
$$= 15(x^2 - 4)(x^2 - 1)$$
$$= 15(x + 2)(x - 2)(x + 1)(x - 1)$$

inc   dec   inc   dec   inc
 +   −   +   −   +
——●————●————●————●————
  −2   −1   1   2
 max   min   max   min

$$f(-2) = -16, f(-1) = -38, f(1) = 38, f(2) = 16$$

$$f''(x) = (15x^4 - 75x^2 + 60)'$$
$$= 60x^3 - 150x$$
$$= 10x(6x^2 - 15)$$
$$= 10x(\sqrt{6}x + \sqrt{15})(\sqrt{6}x - \sqrt{15})$$

down   up   down   up
 −   +   −   +
——————●————●————●————————
    −\sqrt{\frac{15}{6}}   0   \sqrt{\frac{15}{6}}

$$f\left(-\sqrt{\frac{15}{6}}\right) \approx -\frac{25}{7}, f(0) = 0, f\left(\sqrt{\frac{15}{6}}\right) \approx 25.7$$

as $x \to \pm\infty$ the graph of f resembles that of x^5

$(1, 38)$

$-\sqrt{\frac{15}{6}}$

$(2, 16)$

$(-2, -16)$

$\sqrt{\frac{15}{6}}$

$(-1, -38)$

57.
$$\int (x^5 + 3x^2 - x + 1)dx = \frac{x^6}{6} + 3 \cdot \frac{x^3}{3} - \frac{x^2}{2} + x + C = \frac{x^6}{6} + x^3 - \frac{x^2}{2} + x + C$$

58.
$$\int (4x^3 - 2x + 3)dx = 4 \cdot \frac{x^4}{4} - 2 \cdot \frac{x^2}{2} + 3x + C = x^4 - x^2 + 3x + C$$

59. $\int \left(x^4 - 2x^2 + \dfrac{1}{x^3} \right) dx = \int (x^4 - 2x^2 + x^{-3}) dx = \dfrac{x^5}{5} - 2 \cdot \dfrac{x^3}{3} + \dfrac{x^{-2}}{-2} + C = \dfrac{x^5}{5} - \dfrac{2x^3}{3} - \dfrac{1}{2x^2} + C$

60. $\int x^4(x^2 + x - 1) dx = \int (x^6 + x^5 - x^4) dx = \dfrac{x^7}{7} + \dfrac{x^6}{6} - \dfrac{x^5}{5} + C$

61. $\int (x^2 + 5)(x - 3) dx = \int (x^3 - 3x^2 + 5x - 15) dx = \dfrac{x^4}{4} - x^3 + \dfrac{5x^2}{2} - 15x + C$

62. $\int \dfrac{4x^5 + x - 3}{x^3} dx = \int (4x^2 + x^{-2} - 3x^{-3}) \, dx = 4 \cdot \dfrac{x^3}{3} + \dfrac{x^{-1}}{-1} - 3 \cdot \dfrac{x^{-2}}{-2} + C$

$$= \dfrac{4x^3}{3} - \dfrac{1}{x} + \dfrac{3}{2x^2} + C$$

63. $f'(x) = x^2 + 5 \Rightarrow f(x) = \dfrac{x^3}{3} + 5x + C$

since $f(3) = 1$: $1 = \dfrac{3^3}{3} + 5 \cdot 3 + C$

$C = 1 - 9 - 15 = -23$

So: $f(x) = \dfrac{x^3}{3} + 5x - 23$

64. $f'(x) = 4x^3 + 2x \Rightarrow f(x) = x^4 + x^2 + C$

since $f(1) = 5$: $5 = 1 + 1 + C$

$C = 3$

So: $f(x) = x^4 + x^2 + 3$

65. $\displaystyle\int_0^1 (3x + 4) dx = \left. \dfrac{3x^4}{4} + 4x \right|_0^1 = \left(\dfrac{3}{4} + 4 \right) - (0) = \dfrac{19}{4}$

66. $\displaystyle\int_1^2 (4x^3 - 3x^2 + 1) dx = \left. x^4 - x^3 + x \right|_1^2 = (16 - 8 + 2) - (1 - 1 + 1) = 9$

67. $\displaystyle\int_1^2 \left(5x - \dfrac{1}{x^2} \right) dx = \int_1^2 (5x - x^{-2}) dx = \left. \dfrac{5x^2}{2} - \dfrac{x^{-1}}{-1} \right|_1^2 = \left. \dfrac{5x^2}{2} + \dfrac{1}{x} \right|_1^2 = \left(10 + \dfrac{1}{2} \right) - \left(\dfrac{5}{2} + 1 \right) = 7$

68. $\displaystyle\int_0^1 (3x + 4)(2x + 1) dx = \int_0^1 (6x^2 + 11x + 4) dx = \left. 2x^3 + \dfrac{11x^2}{2} + 4x \right|_0^1 = 2 + \dfrac{11}{2} + 4 = \dfrac{23}{2}$

69. $\displaystyle\int_1^2 \dfrac{3x + 4}{x^3} dx = \int_1^2 (3x^{-2} + 4x^{-3}) dx = \left. \dfrac{3x^{-1}}{-1} + \dfrac{4x^{-2}}{-2} \right|_1^2$

$$= \left. -\dfrac{3}{x} - \dfrac{2}{x^2} \right|_1^2 = \left(-\dfrac{3}{2} - \dfrac{2}{4} \right) - (-3 - 2) = 3$$

70. $\displaystyle\int_{-2}^{-1}\frac{x^5+x^3-3}{x^2}dx = \int_{-2}^{-1}(x^3+x-3x^{-2})\,dx = \frac{x^4}{4}+\frac{x^2}{2}-3\frac{x^{-1}}{-1}\Big|_{-2}^{-1}$

$$= \frac{x^4}{4}+\frac{x^2}{2}+\frac{3}{x}\Big|_{-2}^{-1} = \left(\frac{1}{4}+\frac{1}{2}-3\right)-\left(4+2-\frac{3}{2}\right) = -\frac{27}{4}$$

71. $f(x) = -x^2+4 = -(x^2-4) = -(x+2)(x-2)$

SIGN $f(x)$: $\underset{-2}{\bullet}\ +\ \underset{2}{\bullet}\ -$ graph lies above the x-axis

Area $= \displaystyle\int_{-2}^{2}(-x^2+4)\,dx = -\frac{x^3}{3}+4x\Big|_{-2}^{2}$

$$= \left(-\frac{8}{3}+8\right)-\left(\frac{8}{3}-8\right) = \frac{32}{3}$$

72. $f(x) = 2x^3-2x^2 = 2x^2(x-1)$

SIGN $f(x)$: $\underset{0}{\overset{n}{\bullet}}\ -\ \underset{1}{\overset{c}{\bullet}}\ +\ {}_{2}$ graph lies above the x-axis

Area $= \displaystyle\int_{1}^{2}(2x^3-2x^2)\,dx = \frac{x^4}{2}-\frac{2x^3}{3}\Big|_{1}^{2}$

$$= \left(8-\frac{16}{3}\right)-\left(\frac{1}{2}-\frac{2}{3}\right) = \frac{17}{6}$$

73. The words "Determine those values of x (if any) where the slope of the tangent line to the graph of the function $f(x) = x^3$ is twice the rate of change of the function $g(x) = 2x^3-3x$" are just asking us to solve the equation $f'(x) = 2g'(x)$. Let's do it:

$$3x^2 = 2(6x^2-3)$$
$$3x^2 = 12x^2-6$$
$$x^2 = 4x^2-2 \Rightarrow 3x^2-2 = 0 \Rightarrow (\sqrt{3}x+\sqrt{2})(\sqrt{3}x-\sqrt{2}) = 0 \Rightarrow x = \pm\sqrt{\frac{2}{3}}$$

74. We need to solve the inequality $f'(x) > g'(x)$. Let's do it:

$$(x^3)' > (2x^3-3x)'$$
$$3x^2 > 6x^2-3 \Rightarrow 3x^2-3 < 0 \Rightarrow 3(x^2-1) < 0 \Rightarrow 3(x+1)(x-1) < 0:$$

$\underset{-1}{\overset{c}{+\ \bullet}}\ -\ \underset{1}{\overset{c}{\bullet\ +}}$

$\longrightarrow (-1,1)$

75. $P(x) = R(x) - C(x) = 110x - \left(800 + 95x + \dfrac{x^2}{20}\right) = -\dfrac{x^2}{20} + 15x - 800$.

$P'(x) = -\dfrac{x}{10} + 15 = -\dfrac{1}{10}(x - 150)$.　　SIGN $P'(x)$:

$$\begin{array}{c} \underline{\qquad\quad + \quad\;\; \overset{\text{c}}{\bullet} \quad\; - \qquad\quad} \\ \qquad\quad 150 \\ \text{max} \end{array}$$

Conclusion: Maximum profit is realized at a monthly production of 150 units.

76. $P(x) = R(x) - C(x) = \left(60x - \dfrac{x^2}{10}\right) - (100 + 50x) = -\dfrac{x^2}{10} + 10x - 100$.

$P'(x) = -\dfrac{x}{5} + 10 = -\dfrac{1}{5}(x - 50)$.　　SIGN $P'(x)$:

$$\begin{array}{c} \underline{\qquad\quad + \quad\;\; \overset{\text{c}}{\bullet} \quad\; - \qquad\quad} \\ \qquad\quad 50 \\ \text{max} \end{array}$$

Conclusion: Maximum profit is realized at a monthly production of 50 units.

77. $P(x) = R(x) - C(x) = xp(x) - (1000 + 50x)$

$$= x\left(100 - \dfrac{x}{10}\right) - (1000 + 50x) = -\dfrac{x^2}{10} + 50x - 1000$$

$P'(x) = -\dfrac{x}{5} + 50 = -\dfrac{1}{5}(x - 250)$.　　SIGN $P'(x)$:

$$\begin{array}{c} \underline{\qquad\quad + \quad\;\; \overset{\text{c}}{\bullet} \quad\; - \qquad\quad} \\ \qquad\quad 250 \\ \text{max} \end{array}$$

Conclusion: Maximum profit is realized at a weakly production of 250 units.

78. $P(x) = R(x) - C(x) = xp(x) - (15x + 7500)$

$$= x\left(100 - \dfrac{x}{10}\right) - (15x + 7500) = -\dfrac{x^2}{10} + 85x - 7500$$

$P'(x) = -\dfrac{x}{5} + 85 = -\dfrac{1}{5}(x - 425)$.　　SIGN $P'(x)$:

$$\begin{array}{c} \underline{\qquad\quad + \quad\;\; \overset{\text{c}}{\bullet} \quad\; - \qquad\quad} \\ \qquad\quad 425 \\ \text{max} \end{array}$$

To maximize profit, the company should charge $p(425) = 100 - \dfrac{425}{10} = \$57.50$ per unit.

79.

$\$1200$ can sell 300
$\$(1200 - 20x)$ can sell $300 + 15x$

$$R(x) = \overset{\text{price per unit}}{(1200 - 20x)}\,\overset{\text{number of units}}{(300 + 15x)}$$
$$= -300x^2 + 12000x + 360000$$

$R'(x) = -600x + 12000 = -60(x - 20)$.　　SIGN $R'(x)$:

$$\begin{array}{c} \underline{\qquad\quad + \quad\;\; \overset{c}{\bullet} \quad\; - \qquad\quad} \\ \qquad\quad 20 \\ \text{max} \end{array}$$

To maximize revenue, the company should charge $\$(1200 - 20 \cdot 20) = \$800$ per unit.

80.

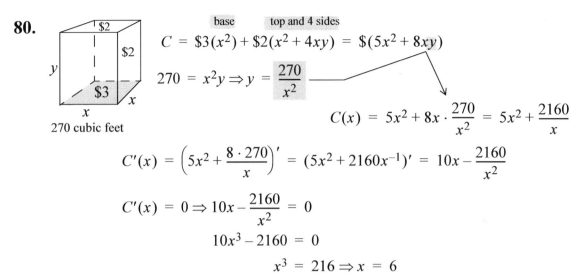

$$C = \$3(x^2) + \$2(x^2 + 4xy) = \$(5x^2 + 8xy)$$

base   top and 4 sides

$$270 = x^2 y \Rightarrow y = \frac{270}{x^2}$$

$$C(x) = 5x^2 + 8x \cdot \frac{270}{x^2} = 5x^2 + \frac{2160}{x}$$

$$C'(x) = \left(5x^2 + \frac{8 \cdot 270}{x}\right)' = (5x^2 + 2160x^{-1})' = 10x - \frac{2160}{x^2}$$

$$C'(x) = 0 \Rightarrow 10x - \frac{2160}{x^2} = 0$$

$$10x^3 - 2160 = 0$$

$$x^3 = 216 \Rightarrow x = 6$$

Realizing that there is no maximum cost and that a minimum cost must exist, we conclude that the minimum material cost is $C(6) = 5 \cdot 6^2 + \dfrac{2160}{6} = \$540$.

81.

$$A = 2xy = 2x(3 - x^2) = 6x - 2x^3$$

$$A'(x) = (6x - 2x^3)' = 6 - 6x^2 = 6(1 + x)(1 - x)$$

SIGN $A'(x)$: $-$ $\overset{c}{\underset{-1}{\bullet}}$ $+$ $\overset{c}{\underset{\underset{\text{max}}{1}}{\bullet}}$ $-$

Maximum area: $A(1) = 6 \cdot 1 - (2 \cdot 1^3) = 4$

82.

perimeter: P

$$A = xy \qquad\qquad A(x) = x\left(\frac{P}{2} - x\right) = \frac{P}{2}x - x^2$$

$$2x + 2y = P \Rightarrow 2y = P - 2x \Rightarrow y = \frac{P}{2} - x$$

$$A'(x) = \frac{P}{2} - 2x = \frac{P}{2}\left(1 - \frac{4}{P}x\right)$$

SIGN $A'(x)$: $+$ $\overset{c}{\underset{\underset{\text{max}}{\frac{P}{4}}}{\bullet}}$ $-$ $\Rightarrow$ maximum area when $x = \dfrac{P}{4}$ $\leftarrow---\urcorner$

and $y = \dfrac{P}{2} - \dfrac{P}{4} = \dfrac{P}{4}$ a square

83. Turning to Theorem 2.16, pave 197, we have:

$$(1): \ v(t) = -32t + 48 \quad \text{and} \quad (2): \ s(t) = -16t^2 + 48t + 64$$

Setting $s(t) = 0$ in (2) we solve for t (the instant when the stone hits the water):

$$-16t^2 + 48t + 64 = 0$$
$$t^2 - 3t + 4 = 0$$
$$(t-4)(t+1) = 0 \Rightarrow t = 4, \text{\rule{1.5cm}{0.3cm}}$$

Substituting 4 for t in (1) we find the impact velocity: $v(4) = -32 \cdot 4 + 48 = -80\dfrac{\text{ft}}{\text{sec}}$ (the negative sign indicates that the stone is moving in a downward direction). Since speed is the magnitude of velocity, we conclude that the stone hits the water at 80 feet per second.

84. Let H denote the number of hours it will take for the tank to empty. Turning to Theorem 2.20, page 208, we have:

$$\int_0^H (7 - 2t)\,dt = 10$$

$$7t - t^2 \Big|_0^H = 10 \Rightarrow 7H - H^2 = 10 \Rightarrow H^2 - 7H + 10 = 0$$
$$(H-5)(H-2) = 0 \Rightarrow H = 5, H = 2$$

Whoa! Will the tank empty twice — once after 5 hours and another tine after 2 hours? No. The math is oblivious to the reality of the problem. Mathematically, the given rate of change $7 - 2t$ gallons per hour will be negative after $t = \dfrac{7}{2}$, which would indicate that water is leaking out at a negative rate, which is to say that water is being poured into the tank. Bottom line: it will take the tank two hour to empty.